Volume 66

Advances
in
Genetics

Advances in Genetics, Volume 66

Serial Editors

Theodore Friedmann
University of California at San Diego, School of Medicine, USA

Jay C. Dunlap
Dartmouth Medical School, Hanover, NH, USA

Stephen F. Goodwin
University of Oxford, Oxford, UK

Volume 66

Advances in Genetics

Edited by

Theodore Friedmann

Department of Pediatrics
University of California at San Diego
School of Medicine, CA, USA

Jay C. Dunlap

Department of Genetics
Dartmouth Medical School
Hanover, NH, USA

Stephen F. Goodwin

Department of Physiology, Anatomy and Genetics
University of Oxford
Oxford, United Kingdom

AMSTERDAM • BOSTON • HEIDELBERG • LONDON
NEW YORK • OXFORD • PARIS • SAN DIEGO
SAN FRANCISCO • SINGAPORE • SYDNEY • TOKYO

ELSEVIER Academic Press is an imprint of Elsevier

Academic Press is an imprint of Elsevier

525 B Street, Suite 1900, San Diego, CA 92101-4495, USA
30 Corporate Drive, Suite 400, Burlington, MA 01803, USA
32 Jamestown Road, London, NW1 7BY, UK
Radarweg 29, POBox 211, 1000 AE Amsterdam, The Netherlands

First edition 2009

Notice

No responsibility is assumed by the publisher for any injury and/or damage to persons or
property as a matter of products liability, negligence or otherwise, or from any use or operation
of any methods, products, instructions or ideas contained in the material herein. Because of rapid
advances in the medical sciences, in particular, independent verification of diagnoses and drug
dosages should be made.

ISBN: 978-0-12-374831-7
ISSN: 0065-2660

For information on all Academic Press publications
visit our website at elsevierdirect.com

Printed and bound in USA

09   10   11   12   10   9   8   7   6   5   4   3   2   1

Working together to grow
libraries in developing countries

www.elsevier.com | www.bookaid.org | www.sabre.org

ELSEVIER   BOOK AID International   Sabre Foundation

Contents

Contributors

Numbers in parentheses indicate the pages on which the authors' contributions begin.

Francisco E. Baralle (1) International Centre for Genetic Engineering and Biotechnology (ICGEB), Trieste, Italy

Emanuele Buratti (1) International Centre for Genetic Engineering and Biotechnology (ICGEB), Trieste, Italy

Jinnie M. Garrett (35) Department of Biology, Hamilton College, Clinton, New York, NY, USA

Joseph V. Gray (61) Molecular Genetics and Integrative & Systems Biology, Faculty of Biomedical and Life Sciences, University of Glasgow, Glasgow, United Kingdom

Sue A. Krause (61) Molecular Genetics and Integrative & Systems Biology, Faculty of Biomedical and Life Sciences, University of Glasgow, Glasgow, United Kingdom

Kathleen L. Triman (35) Department of Biology, Franklin and Marshall College, Lancaster, PA, USA

1

The Molecular Links Between TDP-43 Dysfunction and Neurodegeneration

Emanuele Buratti and Francisco E. Baralle
International Centre for Genetic Engineering and Biotechnology (ICGEB),
Trieste, Italy

ABSTRACT

TDP-43 nuclear protein is involved in several major neurodegenerative diseases that include frontotemporal lobar degeneration with ubiquitin (FTLD-U) bodies and amyotrophic lateral sclerosis (ALS). As a consequence, the role played by this protein in both normal and diseased cellular metabolism has come under very close scrutiny. In the neuronal tissues of affected individuals TDP-43 under-goes aberrant localization to the cytoplasm to form insoluble aggregates. Further-more, it is subject to degradation, ubiquitination, and phosphorylation. Understanding the pathways that lead to these changes will be crucial to define

Advances in Genetics, Vol. 66
0065-2660/09 $35.00
DOI: 10.1016/S0065-2660(09)66001-6

the functional role played by this protein in disease. Several recent biochemical and molecular studies have provided new information regarding the potential physiological consequences of these modifications. Moreover, the discovery of TDP-43 mutations associated with disease in a limited number of cases and the data from existing animal models have strengthened the proposed links between this protein and disease. In this review we will discuss the available data regarding the biochemical and functional changes that transform the wild-type endogenous TDP-43 in its pathological form. Furthermore, we will concentrate on examining the potential pathological mechanisms mediated by TDP-43 in different gain- versus loss-of-function scenarios. In the near future, this knowledge will hopefully increase our knowledge on disease progression and development. Moreover, it will allow the design of innovative therapeutic strategies for these pathologies based on the specific molecular defects causing the disease. © 2009, Elsevier Inc.

I. INTRODUCTION

Nuclear factor TDP-43 is a multifunctional RNA binding protein that has been described to play a role in a great variety of cellular processes such as transcription, pre-mRNA splicing, stability, transport, and translation. In a recent review we have tried to be as comprehensive as possible (Buratti and Baralle, 2008) but the fast pace of research makes it desirable to look at the field again especially with regards to its involvement in neurodegenerative diseases. From a human disease point of view, TDP-43 was first described to participate in the development of monosymptomatic and full forms of cystic fibrosis through its inhibitory effects on the recognition of CFTR exon 9 (Buratti et al., 2001). More recently, its major claim to fame comes from the observation that TDP-43 is the main protein component of the intracellular inclusions found in the neuronal tissues of patients affected by a series of neurodegenerative diseases which include frontotemporal lobar degeneration ubiquitin (FTLD-U), amyotrophic lateral sclerosis (ALS) (Arai et al., 2006; Geser et al., 2009b; Neumann et al., 2006), and Alzheimer disease (Amador-Ortiz et al., 2007; Bigio, 2008; Rohn, 2008). The number of neurodegenerative diseases involved, the distribution/morphology of inclusions in different brain regions, and the associated clinical characteristics have already been reviewed in several recent publications (Bugiani, 2007; Cook et al., 2008; Dickson, 2008; Dickson et al., 2007; Elman et al., 2008; Forman et al., 2007; Kwong et al., 2008; Liscic et al., 2008; Mackenzie and Rademakers, 2007, 2008; Neumann et al., 2007b; Tolnay and Frank, 2007; Wang et al., 2008b). Indeed, the impact of TDP-43 in the neurodegeneration field has been so pervasive that disease nomenclature consensus are currently being modified to better reflect the new clinical and pathological findings originating from recent research (Geser et al., 2009a; Mackenzie et al., 2009).

In this review we will rather concentrate on available data regarding the biochemical and functional changes of TDP-43 that lead to the human pathology. In particular we will focus on the potential pathological mechanisms mediated by TDP-43 mutations (both natural and artificial) and the results obtained so far in simple or existing animal models of disease.

From a classification point of view, TDP-43 is a member of the hnRNP protein family (Krecic and Swanson, 1999), a family that includes many of the most common and powerful splicing modulators known so far, such as hRNP A1/A2, PTB (hnRNP I), and hnRNP H (Martinez-Contreras et al., 2007). A characteristic of many of these factors, however, is the role they play in numerous diverse functions depending on their relative abundance, cellular localization, and the interactions with themselves or other components of the cellular machinery (Carpenter et al., 2006; Dreyfuss et al., 2002; Glisovic et al., 2008). TDP-43 is no exception and Fig. 1.1 shows a schematic diagram of the protein and the best characterized functions reported previously. These include the ability of TDP-43 to regulate splicing of several exons such as CFTR exon 9 (Buratti et al., 2001), ApoAII exon 3 (Arrisi-Mercado et al., 2004; Mercado et al., 2005), and SMN exon 7 (Bose et al., 2008) or its role in regulating transcription of the HIV-1 genome (Ou et al., 1995) and of the SP-10 mouse promoter (Abhyankar et al., 2007; Acharya et al., 2006).

In addition to its role in transcription/splicing TDP-43 may also possess other functions that are still at a very early stage of characterization. These include some very diverse properties such as acting as a neuronal response activity factor and an in vitro mRNA translational repressor (Wang et al., 2008a), or an mRNA stability factor for neurofilaments (Strong et al., 2007).

Finally, it is also important to mention the possibility of some additional properties, such as microRNA processing, that up to this moment are simply based on its association with both the human and mouse microprocessor complexes (Fukuda et al., 2007; Gregory et al., 2004). The list of potential functions is not likely to end here, as according to protein–protein association studies that use high-throughput methodologies several other candidates have been put forward (Wang et al., 2008b). In particular, two proteomics studies (Lehner and Sanderson, 2004; Stelzl et al., 2005) involving yeast two hybrid systems found XRN2 and PM/Scl100 (mRNA decay), ZHX1 (transcriptional repressor), SETDB1 (chromatin remodeling regulator), and NSFL1C and ARF6 (both involved in membrane trafficking) as potential TDP-43 binding partners. In this respect, it is important to note that TDP-43 has been described to be part of RNA granules responsible for trafficking, sequestering, and degrading RNA species (Elvira et al., 2006) and has been observed to colocalize strongly with

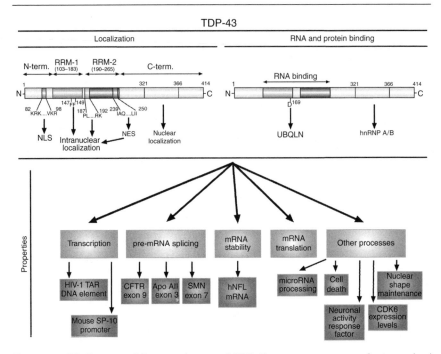

Figure 1.1. *DP-43 protein.* Schematic diagram of TDP-43 sequences important for its nuclear/ cytoplasmic localization (left) and its RNA and protein binding properties (right). The lower part of the diagram reports some of the described and proposed functions of TDP-43 in cellular metabolism.

Staufen (a protein involved in mRNA transport in dendrites), moderately with TIA-1 (a splicing factor) and weakly with XRN1, an exoribonuclease involved in mRNA decay (Moisse *et al.*, 2009). TDP-43 has also been shown to colocalize with wild-type and mutant forms of valosin-containing protein (VCP) in culture cells (Gitcho *et al.*, 2009) and with UBAP1 in the neuronal cytoplasmic inclusions of a familial FTD case (Rollinson *et al.*, 2009). Partial colocalization has also been observed with GW182 and eIF4E (both markers of P-bodies) (Wang *et al.*, 2008a). Finally, in sporadic ALS patient samples TDP-43 has also been found to colocalize with the phosphorylated Smad2/3 factors (pSmad2/3) (Nakamura *et al.*, 2008), which are central mediators of the TGF-beta signal transduction pathway, essential for maintaining the survival of neurons. In this respect, it should also be noted that TDP-43 has been originally suggested to play a central role in organizing the higher order structures of eukaryotic nuclear bodies (Wang *et al.*, 2002). In conclusion, although for a complete list of TDP-43 properties and functions we will certainly have to wait for a long time these observations are already sufficient to support the view that TDP-43 is involved in more than one cellular process, possibly through interactions of different parts of its molecule.

Interestingly, TDP-43 is not the only DNA/RNA binding protein recently found to be involved in ALS. In fact, the recent reports regarding the involvement of mutations in the FUS/TLS protein in a series of patients affected by familial forms of ALS suggests that alterations in DNA/RNA processing mechanisms may be a fertile area for research to further our understanding of neurodegenerative diseases (Kwiatkowski *et al.*, 2009; Vance *et al.*, 2009). Hopefully, this will place us in the best prospective position to design viable therapeutic strategies.

Before tackling this issue, however, it is best to summarize what happens to the predominantly nuclear, wild-type TDP-43 in a pathological setting.

II. MAIN TEXT

As clearly highlighted by the two pioneering studies in this field, the pathological TDP-43 protein analyzed in patients displays some very clear modifications with respect to the wild-type protein (Arai *et al.*, 2006; Neumann *et al.*, 2006). These include an increased cytoplasmic localization under the form of insoluble aggregates, ubiquitination, phosphorylation, and degradation to yield C-terminal fragments (schematically reported in Fig.1.2). In the past 2 years, several advances have been made with regards to explaining these aspects. In the following paragraphs, we will briefly review what is currently known about these aberrant events.

A. Aggregation/nuclear localization

With some exceptions, TDP-43 aberrant aggregation and cellular localization are usually linked with each other. In normal conditions, independent studies agree with the conclusion that TDP-43 is a predominantly a nuclear protein with a very small amount being present in the cytoplasm (Ayala *et al.*, 2008b; Winton *et al.*, 2008a). In rat hippocampal neurons the percentage of cytoplasmic TDP-43 has been recently estimated as 15% (Wang *et al.*, 2008a). The small amount present in the cytoplasm, however, should not be considered simply a cellular background. In fact, using heterokaryon assays it has been noted that TDP-43 is continuously shuttling between the nucleus and the cytoplasm unlike exclusively nuclear factors (Ayala *et al.*, 2008b). This finding suggests that TDP-43 cytoplasmic levels may play a functional role also in normal conditions. On the other hand, a typical hallmark of diseased cells is represented by the almost complete absence of nuclear TDP-43 and staining for this protein can be observed

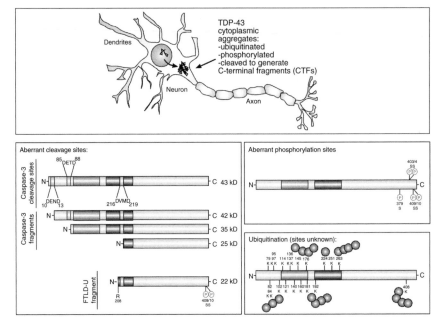

Figure 1.2. *Pathological TDP-43 characteristics.* Characteristic features of pathological TDP-43 in neurodegeneration. The normally nuclear TDP-43 is transported to the cytoplasm of the neuron (upper diagram), it's cleaved by caspases to generate − 35 and − 25 kDa fragments (lower left) or by unknown proteases in the − 22 kDa fragment. The fragments and the full length protein become aberrantly phosphorylated at serine residues 379, 403/404, and 409/410 (right diagram). In addition to all these modifications, it is ubiquitinated in still unknown residues (all lysine residues, which represent the normal substrate for ubiquitination, are shown in the lower right picture).

predominantly in the cytoplasm, where it is often found in insoluble aggregate form. In this respect, it is worthy to point out that TDP-43 aggregations in brain tissues has not been detected in response to metabolic insults such as anoxia or ischemia (Lee *et al.*, 2008). One important general observation concerning aggregation is that TDP-43 tends to be a very "sticky" protein even in normal or slightly altered conditions. In fact, some reports describe aggregation of wild-type TDP-43 in common laboratory culture cell lines or yeast cells (Johnson *et al.*, 2008; Winton *et al.*, 2008a). Moreover, a recent report has also reported the observation that isolated RRM-2 domains can assemble in higher order particularly thermal-stable assemblies in physiological (*in vitro*) conditions (Kuo *et al.*, 2009). As formation of these structures does not occur with the full protein itself this observation suggests that the degradation processes produces TDP-43 fragments with increased tendency to aggregation.

Clearly, a first step to understand the modulation of TDP-43 cytoplasmic localization is to map the sequences that make TDP-43 a predominantly nuclear protein. TDP-43 has been shown to have a bipartite nuclear localization signal (NLS) within residues 82–98 (Winton et al., 2008a) of the N-terminal domain. Deletion/mutation of this sequence causes a prominent nuclear to cytoplasmic shift of TDP-43 localization, a result that has been confirmed independently by other groups (Ayala et al., 2008b; Inukai et al., 2008). Moreover, it has to be noted that mutations in this NLS sequence may well represent a potential risk factor for disease as exemplified by a missense A90V variation found in a FTLD/ALS patient with a family history of dementia that has been associated with an increased TDP-43 presence in the cytoplasm and pathological aggregate formation (Winton et al., 2008b). Another important sequence identified by Winton et al. (2008a) that may contribute to modulate the NLS action is represented by the nuclear export sequence (NES) localized at residues 235–250 of TDP-43. Deletion of this sequence led to the formation of intranuclear inclusions. These NLS and NES sequences, however, are not the only determinants of TDP-43 cellular localization.

Recently, for example, the integrity of the C-terminal tail has also been shown to be very important with regards to the localization of this protein as its progressive deletion results in the formation of large nuclear and cytoplasmic aggregates if the protein is overexpressed (Ayala et al., 2008b). This region can bind other members of the hnRNP A/B family (especially hnRNP A2) (He and Smith, 2007) and is essential for TDP-43 to function as splicing silencer in the CFTR exon 9 and Apo AII exon 3 systems (Ayala et al., 2005; Buratti et al., 2005) (Fig. 1.2). Furthermore, the C-terminal region is also required for the ability of TDP-43 to act as a transcriptional insulator for the mouse SP-10 gene (Abhyankar et al., 2007). However, immunohistochemical studies performed on cytoplasmic TDP-43 inclusions in brain tissues from patients have observed the absence of these hnRNP proteins (Neumann et al., 2007a), suggesting that maintenance of TDP-43 biochemical functionality may well be a factor to keep this protein predominantly nuclear and smoothly shuttling between nucleus and cytoplasm.

The possibility that increased transport from the nucleus to the cytoplasm may represent a physiologic response following neuronal injury has been recently addressed by Moisse et al. (2009). These authors have shown that, in both distally and proximally axotomized mice there is an increased expression and relocalization of TDP-43 from the nucleus to the cytoplasm (which is also reversible), suggesting that this protein may play an early role in neuronal repair. Interestingly, Western blot analysis of TDP-43 from lumbar hemicords did not reveal any difference from controls, also suggesting that cytoplasmic TDP-43 does not necessarily correlate with aberrant modifications. This conclusion is consistent with the results recently obtained in a SH-SY5Y cell model of disease where evident cytoplasmic TDP-43 inclusions of a mutant carrying a deletion in the NLS could be observed only following addition of the proteasome inhibitor MG132 (Nonaka et al., 2008).

Beside nuclear/cytoplasmic distribution, it is now also becoming clear that the localization of TDP-43 within the nuclear compartment is just as tightly regulated. For example, Ayala et al. (2008b) have shown that an extensive deletion in the RRM-1 motif region (residues 106–175) leads to the formation of dot-like structures within the nucleus that are associated with a substantial redistribution of TDP-43 in chromatin fractions. This effect could also be achieved by simply mutating the two key Phe residues (147–149) that have been described as essential for TDP-43 correctly binding to UG-rich RNA (Buratti and Baralle, 2001), suggesting that RNA binding determines TDP-43 intranuclear distribution. Although there are differences in the morphology of the nuclear bodies, the TDP-43 with RRM-1 deleted and the TDP-43 with the Phe 147–149 mutated are identical regarding the chromatin fractions. From this observation it can be inferred that RNA binding may affect TDP-43 mobility from insoluble heterochromatin bound to a looser chromatin interaction or even nucleoplasm solubility. Interestingly, the formation of apparently similar dot-like structures has also been recently observed in a mutant carrying a 187–192 residues deletion, in which TDP-43 also displayed a low level of aberrant phosphorylation (Nonaka et al., 2008).

Taken together, all these data suggest that TDP-43 cellular distribution is a tightly regulated process and that numerous sequences in its primary amino acid structure can contribute to the final picture. Furthermore, it is clear that cellular distribution is not an exclusive function of TDP-43 sequence, but that general cellular localization and aggregation controlling factors play a critical role. For example, it has recently been observed that injection in culture cells of mutant VCP associated with FTLD disease results also in the altered localization of TDP-43 from the nucleus to the cytoplasm, in addition to other effects such as apoptosis, ER-stress, and impairment in cell viability (Gitcho et al., 2009). The observation that VCP and TDP-43 coimmunoprecipitate together not only from transfected cells but also from brain homogenates may point to a functional relationship between these two proteins (Gitcho et al., 2009).

In normal conditions, the integrity of the general cellular trafficking and transport mechanisms of the cell may also play a role in keeping TDP-43 proper cellular distribution. First of all, it is important to note that TDP-43 has been described to be part of RNA granules responsible for trafficking, sequestering, and degrading RNA species (Elvira et al., 2006). More recently, it has been shown that its clearance by autophagic pathways is an additionally important factor in determining its eventual aggregation in the cytoplasm. Evidence for this observation has been first obtained by work performed on the endosomal sorting complexes required for transport (ESCRT). These complexes are important for recognition of ubiquitinated endocytosed integral membrane proteins, their sorting in multivesicular body (MVB), and for subsequent degradation in lysosome/vacuoles. Depletion of ESCRT subunits in HeLa cells was shown to correlate with endogenous TDP-43 accumulating in ubiquitin-positive cytoplasmic aggregates (Filimonenko *et al.*, 2007). Another interesting potential link between TDP-43 accumulation and defects in intracellular transports concern Perry's syndrome, which is characterized by rapidly progressive and fatal autosomal dominant parkinsonism, central hypoventilation, depression, and severe weight loss. In this case, TDP-43 inclusions are also quite distinguished from those observed in FTD/ALS as they occur only in the extrapyramidal system but are absent in the neocortex and motoneurons (Wider *et al.*, 2008). The connection with the transport system has been recently uncovered following the discovery that patients affected by this disease present mutations in the *DCTN1* gene that codes for the large subunit of the dynactin complex p150^{glued} (Farrer *et al.*, 2009). Dynactin is a multiprotein complex that is important for microtubule-based motility and anchoring at centrosomes. Interestingly, dynactin complex proteins have been observed to colocalize with TDP-43 inclusions in some cases, raising the possibility that TDP-43 aggregation may possess a direct connection with dynactin complex disruption.

Finally, it has to be noted that Golgi apparatus fragmentation has been reported to be a common feature of several neurodegenerative diseases including ALS. Interestingly, disruption of the Golgi apparatus in the anterior horn cells of neurons form ALS-affected patients has been shown to correlate with cytoplasmic localization of TDP-43 (Fujita *et al.*, 2008). Considering that Golgi fragmentation seems to be an early event in neurodegeneration it may well be that early disruption of the cellular transport, targeting, and other Golgi-mediated activities results in abnormal TDP-43 localization/aggregation, although further work will be required to exactly determine the degree of Golgi disruption in other neurodegenerative diseases and its pathological connection.

B. Degradation

Increased degradation is a hallmark of TDP-43 in affected neuronal tissues. The initial studies by the groups of V. Lee and T. Arai/M. Hasegawa (Arai *et al.*, 2006; Neumann *et al.*, 2006) have both described the presence of -25 and -35 kDa fragments that are normally absent from normal controls. A recent report has suggested that these fragments derive from the activation of caspase-dependent cleavage and results in redistribution of the fragments in the cytoplasm (Zhang *et al.*, 2007). In keeping with these conclusions, the wild-type TDP-43 contains three well-defined caspase cleavage sites that are predicted to yield fragments of a size compatible with those observed in the patients (highlighted in Fig. 1.2, lower left panel). In particular, caspase-3 activation has been proposed to occur following the downregulation of the *PGRN* gene and considering that mutations in this gene are associated with the occurrence of frontotemporal lobar degeneration (Ahmed *et al.*, 2007; Pickering-Brown, 2007) this observation has for the first time suggested a functional link between progranulin dysfunction in FTLD cases and the observed TDP-43 cleavage. The connection with *PGRN* and TDP-43 degradation, however, could not be replicated in a subsequent study and further work will be required to clarify this point (Shankaran *et al.*, 2008). It should also be noted that −25 and −35 kDa fragments may not be the only degradation products of TDP-43 inside the affected cells. In fact, a recent analysis of a −22 kDa fragment from brain tissue of an FTLD-U patient has demonstrated that this fragment occurs in correspondence to residue 208 (Fig. 1.2, lower left panel) (Igaz *et al.*, 2009). Interestingly, transfection of this TDP-43 fragment in culture cell lines resulted in cytoplasmic localization and aggregate formation. Furthermore, it also resulted in the modification of splicing profiles as determined using an artificial minigene carrying CFTR exon 9 sequences (Igaz *et al.*, 2009), possibly through the sequestering of hnRNP proteins or wild-type TDP-43 in the cytoplasm. Although in a very artificial setting, this finding suggests that degradation of TDP-43 may originate biologically active peptides which can aberrantly modify normal TDP-43 regulated events.

Another issue that remains to be determined is whether degradation of TDP-43 is present in all patient tissues or just in the affected brain tissues. At present, no systematic search has been performed in this direction and most data available concern two of the "easiest" choices to obtain sample tissues, human blood in its various fractions and cerebrospinal fluid (CSF). With regards to human blood, the TDP-43 protein analyzed in Western blot from plasma samples from FTD and AD patients did not display any aberrant mobility or presence of degradation products (although it could be argued that, as it was immunocaptured prior to analysis, eventual fragments/modified TDP-43s may have been lost following this procedure) (Foulds *et al.*, 2008). Similarly, in CSF of patients affected by FTLD and ALS the use of a C-terminus-specific antibody against TDP-43 could detect it as a 45 kDa bands that did not display any signs of pathological alteration (Steinacker *et al.*, 2008). In this respect, a 43 kDa band without any sign of degradation was also detected using western blot in CSF samples from sporadic ALS patients (Kasai *et al.*, 2009) and also in skeletal muscle biopsies from a cohort of ALS patients (Sorarù *et al.*, 2009).

On the other hand, lymphoblastoid cell lines derived from ALS patients carrying disease-associated point mutations displayed the presence of increased TDP-43 degradation with respect to controls in the predominant form of −28 kDa degradation bands, especially following proteasome inhibition by MG132 (Kabashi *et al.*, 2008). In agreement with this, the presence of similar degradation bands has been confirmed also in a recent study where deep-frozen lymphocyte preparations were available from patients carrying missense TDP-43 substitutions (Corrado *et al.*, 2009).

Perhaps even more importantly, all these studies have started to address the crucial point of comparing the antibody preparations currently used in TDP-43 analysis. In fact, the TDP-43 in the CSF could be detected only by antibodies raised against the C-terminal portion of the protein but not by antibodies raised against the N-terminal portion of TDP-43 (Steinacker et al., 2008). This differential reactivity of N-terminal versus C-terminal epitopes has also been highlighted in a study which has shown that TDP-43 inclusions in cortical regions are efficiently labeled with C-terminal antibodies but not with N-terminal antibodies (Igaz et al., 2008). This is in contrast to inclusions in motor neurons in spinal cord where the same N- and C-terminally specific antibodies displayed similar immunoreactivities. These findings raise the interesting suggestion that protein composition of TDP-43 inclusions might differ also depending on their location, an eventuality with completely unknown pathophysiological significance. Specific antibodies may also be useful to perform a reappraisal of previous data. For example, Rohn (2008) recently demonstrated that an antibody specifically designed to detect the -25 kDa caspase-cleaved fragments revealed caspase-degraded TDP-43 as a major disease component in AD brains.

It is therefore clear that the choice of antibodies used in this kind of studies must be evaluated with care and some effort should be spared to accurately determine their target specificities. For example, it has been established that the commercially available mouse monoclonal antibody recognizes the human sequence spanning residues 205–222 and would thus be unable to detect any C-terminal fragments (Zhang et al., 2008). The importance of antibody choice will be further discussed in the section dealing with TDP-43 aberrant phosphorylation.

C. Phosphorylation

Aberrant phosphorylation of TDP-43 was first observed in the initial study by V. Lee's group, (Neumann et al., 2006) where a 45 kDa band corresponding to TDP-43 could be detected in addition to the traditional 43 kDa band corresponding to the molecular weight of normal TDP-43. In their study, Neumann et al. showed that presence of this 45 kDa band was abolished following dephosphorylation of the urea-extracted protein fractions from FTLD-U brains.

The positions of at least some of the aberrantly phosphorylated TDP-43 residues have been recently determined by Hasegawa et al. (2008), who prepared polyclonal antibodies against 36 out of the 64 potentially phosphorylation sites of TDP-43 (represented by 41 Serines, 15 Threonines, and 8 Tyrosines) and used them to screen FTLD-U and ALS patient samples. Using this experimental approach, Hasegawa et al. found that five Serine residues localized at the C-terminal extremity of TDP-43 were specifically phosphorylated (S379, S403,

S404, S409, S410) (summarized in Fig. 1.2). Furthermore, at least *in vitro*, the kinase responsible for this phosphorylation was identified in this study as Casein Kinase 1 or 2 (CK1/2).

Most importantly, the use of monoclonal antibodies engineered to recognize phosphorylated serines S409/S410 showed that these residues undergo aberrant phosphorylation in the inclusions of all familiar and sporadic forms of FTLD-U (carrying mutations in progranulin, VCP, and linkage to chromosome 9p) and ALS (Neumann *et al.*, 2009). In addition, by comparing the levels of staining of "preinclusions" by these antibodies with antiubiquitin antibodies it has also been possible to determine that phosphorylation precedes ubiquitination (Neumann *et al.*, 2009). More recently, using phosphorylation sensitive antibodies it has been shown that the presence of aberrantly phosphorylated/processed TDP-43 in Alzheimer disease and dementia with Lewy bodies is also more frequently present with respect to previous estimates performed using commercial anti-TDP-43 antibodies (Arai *et al.*, 2009). For these reasons, there is no doubt that in the near future the use of phosphorylation-specific antibodies against TDP-43 will represent extremely powerful tools in the diagnostic and basic research field.

The question that still remains open regards whether phosphorylation may also occur in endogenous TDP-43 in normal conditions. Incubating recombinant TDP-43 in total cellular extract from cultured cell lines in the presence of γP32-ATP, followed by immunoprecipitation, has failed to uncover any phosphorylated protein (E. Buratti, F. E. Baralle, unpublished data), suggesting that phosphorylation may be specific of disease (at least within current detection limits). This statement is supported by the extensive analysis of normal rat/mouse tissues (including brains in different developmental stages) which have failed to detect phosphorylated S409/410 TDP-43 (Inukai *et al.*, 2008). Moreover, phosphorylation-specific antibodies have revealed no physiological nuclear TDP-43 staining by immunohistochemistry in cultured cells, mouse, and human brain homogenates (Neumann *et al.*, 2009).

Intriguingly, however, there is the observation that normal endogenous TDP-43 is involved in regulating cellular phosphorylation processes. In fact, TDP-43 has been shown to be a major regulator of the protein and transcript levels of cyclin-dependent kinase 6 (Cdk6) and other factors that control cell proliferation, which significantly increase following its knockout from normal cells (Ayala *et al.*, 2008a).

D. Ubiquitination

The formation of protein aggregates containing ubiquitin, a small protein consisting of 76 amino acids, has long been known to be associated with a great number of neurodegenerative diseases such as FTLD-U itself, Parkinson,

Alzheimer, and Huntington. The ubiquitin proteasome system (UPS) is the major eukaryotic protein degradation pathway. It involves an enzymatic cascade which includes ubiquitin-activating enzymes (E1), ubiquitin conjugating enzymes (E2), and ubiquitin ligases (E3). This system is known to play a very important role in a myriad of processes such as DNA repair, embryogenesis, the regulation of transcription, and apoptosis. Not surprisingly, it plays a very important role in normal neuronal development, as reviewed recently by Segref and Hoppe (2009). At present, the residues involved in TDP-43 ubiquitination are still unknown although they presumably occur in correspondence of lysine residues in TDP-43 (indicated in Fig. 1.2).

A direct link between TDP-43 and the ubiquitin pathway has already been reported by Kim et al. (2008). Using a yeast two-hybrid assay they have demonstrated the interaction between TDP-43 and the proteasome targeting factor, Ubiquilin-1 (UBQLN). In normal conditions, binding of UBQLN to ubiquitylated TDP-43 stimulates aggregation and may represent a cytoprotective response by sequestering it in autophagosomes. From the point of view of the disease-causing effect, it is particularly interesting to note that the TDP-43 mutant (D169G) found in ALS patients is unable to bind UBQLN. This may provide a molecular explanation for the effect of this mutation as the mutated protein would be unable to be properly sequestered and thus might freely accumulate in the cytoplasm leading to aberrant aggregation/degradation. In addition, as previously mentioned, in a familial case of FTLD there is co-immunolocalization of TDP-43 and a disease-associated haplotype of ubiquitin-associated protein 1 (UBAP1) in neuronal cytoplasmic inclusions. This observation further strengthens the potential association between TDP-43 ubiquitination pathway and the newly identified role of UBAP1 as a risk factor for frontotemporal lobar degeneration (Rollinson et al., 2009).

E. Mutations

The search for TDP-43 mutations in affected patients has been quite intensive in an attempt to establish a clear molecular connection with disease and thus rule out the possibility that TDP-43 aggregates might be just a "pathological curiosity" (Rothstein, 2007). Initial studies in a FTLD patient cohort failed to find an increased genetic risk factor for FTLD when analyzing common TDP-43 variations, especially with regards to single-nucleotide-polymorphism (SNP) (Rollinson et al., 2007), a conclusion that has also been confirmed in a more recent study (Schumacher et al., 2009). Finally, TARDBP sequencing of a Canadian Population of 125 patients affected by Parkinson disease also failed to detect any mutations in this gene (Kabashi et al., 2009).

Nonetheless, approximately 2–3% of patients affected by sporadic and familial forms of ALS carry specific missense mutations in the C-terminus of TDP-43 (Corrado et al., 2009; Daoud et al., 2008; Del Bo et al., 2009; Gitcho et al., 2008; Kabashi et al., 2008; Kuhnlein et al., 2008; Lemmens et al., 2009; Rutherford et al., 2008; Sreedharan et al., 2008; Van Deerlin et al., 2008; Yokoseki et al., 2008) with some of these studies recently reviewed by Banks et al. (2008). This percentage may be different depending on the populations analyzed as similar screening studies performed on a mixed Belgian FTD/ALS population (Gijselinck et al., 2007) or on a mixed European-descent North American ALS population (Guerreiro et al., 2008) failed to find any copy number variation or amino acidic mutation in the TDP-43 gene.

The type of mutations described so far together with their functional effects (when known) or predicted impact on protein functionality/properties has been summarized in Table 1.1. It should be noted that to this date these analyses have involved the screening of approximately 3600 FTD/ALS patients and 7900 controls. Furthermore, none of the disease-associated mutations found in any screening study has been found in control cases from other screening studies.

All mutations described so far are missense substitutions with the exception of truncation mutant Y374X. The spatial distribution of these mutations in the C-terminal tail of TDP-43 together with the number of patients in which they have been detected is reported in Fig. 1.3. As shown in this figure, all 29 different mutations are localized across the entire C-terminal sequence of TDP-43 with just a tendency of clustering in specific positions. This fact, together with the observation that the only missense substitutions not associated with disease (D65E and A90V) fall outside the C-terminal tail suggests that any alteration in residues 296–414 of TDP-43 may be potentially linked with or predisposing to disease.

Presently, with the exception of D169G, that has been proposed to inhibit TDP-43 interaction with UBQLN (Kim et al., 2008), there is very limited information with regards to which specific activity of TDP-43 the other mutations may be affecting/introducing. For the moment, M337V/Q331K has been associated with neurotoxicity (Sreedharan et al., 2008) and G348C/R361S/N390D (Kabashi et al., 2008) and S393L/A382T (Corrado et al., 2009) with increased degradation. In none of these cases, however, a direct link with some functional characteristic has been established.

It should also be noted that the C-terminal tail of TDP-43 does not seem to possess a higher degree of conservation with respect to other regions of the protein (data not shown). In addition, many supposedly pathogenic substitutions in humans are found in the wt gene of other species, where they do not seem to alter their respective TDP-43 protein functions. Taken together, these observations suggest that the importance of these mutations in human disease

Table 1.1. Review of TARDBP Mutations and the Observed/Potential Consequences on Protein Functionality

Mutation	Disease	Experimental consequences of the mutation	Potential consequences based on *in silico* analyses	References
p.D169G	SALS	Reduces binding to UBQLN factor	n.a.	Kabashi *et al.* (2008)
p.N267S	SALS	n.a.	Probable increased phosphorylation	Corrado *et al.* (2009)
p.G287S	SALS	n.a.	n.a.	Corrado *et al.* (2009) and Kabashi *et al.* (2008)
p.G290A	FALS	n.a.	n.a.	Van Deerlin *et al.* (2008)
p.G294V	FALS/SALS	n.a.	n.a.	Corrado *et al.* (2009) and Del Bo *et al.* (2009)
p.G294A	SALS	n.a.	n.a.	Sreedharan *et al.* (2008)
p.G295S	SALS	n.a.	Probable increased phosphorylation	Corrado *et al.* (2009) and Del Bo *et al.* (2009)
p.G295R	SALS	n.a.	n.a.	Corrado *et al.* (2009)
p.G298S	FALS	n.a.	n.a.	Van Deerlin *et al.* (2008)
p.M311V	FALS	n.a.	n.a.	Lemmens *et al.* (2009)
p.A315T	FALS	n.a.	n.a.	Gitcho *et al.* (2008) and Kabashi *et al.* (2008)
p.Q331K	SALS	Causes neurotoxicity when injected in chicken embryos	Probable increased phosphorylation	Sreedharan *et al.* (2008)
p.S332N	FALS	n.a.	n.a.	Corrado *et al.* (2009)
p.G335D	SALS	n.a.	n.a.	Corrado *et al.* (2009)
p.M337V	FALS	Causes neurotoxicity when injected in chicken embryos and increased degradation in lymphoblastoid cell line	n.a.	Corrado *et al.* (2009); Rutherford *et al.* (2008); and Sreedharan *et al.* (2008)
p.Q343R	FALS	Probable increase in fragment production	n.a.	Yokoseki *et al.* (2008)

(Continues)

Table 1.1. (*Continued*)

Mutation	Disease	Experimental consequences of the mutation	Potential consequences based on *in silico* analyses	References
p.N345K	FALS	Increased degradation in lympho-blastoid cell line	n.a.	Rutherford *et al.* (2008)
p.G348C	FALS/SALS	Increased degradation in lympho-blastoid cell line	n.a.	Daoud *et al.* (2008); Del Bo *et al.* (2009); Kabashi *et al.* (2008); and Kuhnlein *et al.* (2008)
p.N352S	FALS	n.a.	Probable increased phosphorylation	Kuhnlein *et al.* (2008)
p.R361S	SALS	Increased degradation in lympho-blastoid cell line	n.a.	Kabashi *et al.* (2008)
p.P363A	SALS	n.a.	n.a.	Daoud *et al.* (2008)
p.Y374X	SALS	n.a.	n.a.	Daoud *et al.* (2008)
p.S379C	SALS	n.a.	n.a.	Corrado *et al.* (2009)
p.S379P	FALS	n.a.	n.a.	Corrado *et al.* (2009)
p.A382T	FALS/SALS	n.a.	n.a.	Corrado *et al.* (2009); Del Bo *et al.* (2009); and Kabashi *et al.* (2008)
p.A382P	SALS	n.a.	n.a.	Daoud *et al.* (2008)
p.I383V	FALS	Increased degradation in lympho-blastoid cell line	n.a.	Rutherford *et al.* (2008)
p.N390D	SALS	Increased degradation in lympho-blastoid cell line	Probable increased phosphorylation	Kabashi *et al.* (2008)
p.N390S	SALS	n.a.	Probable increased phosphorylation	Kabashi *et al.* (2008)
p.S393L	SALS	n.a.	n.a.	Corrado *et al.* (2009)

FALS, Familial ALS; SALS, sporadic ALS, n.a., not available.

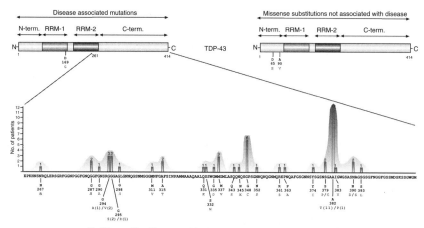

Figure 1.3. *TDP-43 disease-associated mutations.* Schematic diagram of the C-terminal tail of TDP-43 and of the missense/truncation mutations associated with disease (with the exception of D169G that is reported in the upper left diagram). The upper right diagram also indicates the two missense variations identified in screening studies that cannot be associated with disease owing to their presence in controls or both in affected and control individuals (in the case of A90V). The bar chart indicates the number or patients described carrying each of the identified mutation (in case of mutations found in different affected family members only one patient for family was computed).

should not be overemphasized. In fact, there still exists the possibility that rather than being directly pathogenic most of these changes observed in the C-terminal tail are simply predisposing to alterations: for example, by slightly altering shuttling properties or interactions that would eventually lead to easier accumulation in the cytoplasm and consequent knockout in the nucleus.

III. EARLY LESSONS FROM ANIMAL MODELS

The field of animal disease models dealing with TDP-43 is still in its infancy. Since the discovery of TDP-43 involvement in FTD/ALS too little time has passed to engineer TDP-43 knockout/knock-in models in higher animals. Nonetheless, some information is already available with regards to the analysis of TDP-43 expression in simpler organisms and in respect to already established animal models of ALS (Kato, 2008). The results of these studies are summarized in Table 1.2. For example, TDP-43 expression has been studied in a yeast system already used for the study of other potential neurodegenerative proteins such as alfa-synuclein and polyglutamine. The data show that human TDP-43 was

Table 1.2. Cellular and Animal Neurodegenerative Models Investigated for TDP-43 Involvement

Organism	Model name	Disease model	Gene affected	Description and results	References
Saccharomyces cerevisiae (Yeast)	n.a	n.a	n.a.	Overexpression of TDP-43 or C-terminal fragments containing the RRMs in yeast cells is associated with cellular toxicity and cytoplasmic accumulation	Johnson et al. (2008)
Mus Musculus (Mouse)	SOD1 mice	ALS	SOD1 gene carrying G37R, G85R, and G93A missense mutations	Mice did not display any mislocalization of TDP-43 in the cytoplasm of motor neurons nor were found associated with ubiquitinated inclusions	Robertson et al. (2007) and Turner et al. (2008)
Mus Musculus (Mouse)	"Wobbler" mice	Spinal cord disorders	Vps54 gene carrying a L967Q missense mutation	Mice display elevated levels of TDP-43 mRNA and protein in spinal cord with respect to controls. Furthermore, TDP-43 in mice affected by disease is sequestered in the cytoplasm of spinal cord cells	Dennis and Citron (2009)
Mus Musculus (Mouse)	Smn-/- and SMN2; SMNΔ7	Spinal Muscolar Atrophy	Both Smn genes are knocked out	Reduction in SMN protein was not accompanied by alterations in the mRNA or protein levels of TDP-43. Furthermore, no qualitative or quantitative changes in TDP-43 distribution were observed within nuclei	Turner et al. (2008)

Rattus norvegicus (rat)	FTLD-ubiquitin; ALS	n.a.	n.a.	Using an adenovirus associated virus (AAV) wild-type human TDP-43 was injected in the nucleus and cytoplasm of rat midbrain neurons. Exogenous TDP-43 expression resulted in gliosis, neuronal loss, and progressive motor dysfunctions in the injected animals	Tatom et al. (2009)
Drosophila melanogaster (Fruitfly)	n.a.	n.a.	Deletion of the TBPH gene (homologue of TARDBP)	Flies lacking TBPH appear externally normal but present deficient locomotor behaviors, reduced life span, and anatomical defects at the neuromuscular junctions. The phenotype can be rescued by expression of human TDP-43 in the motoneurons of TBPH-deficient Drosophila	Feiguin et al. (2009)

n.a., Not applicable.

normally localized in the nucleus but that upon overexpression formed cytoplasmic inclusions which were highly toxic provided that the aggregating
TDP-43 mutants contained the RRM domains, broadly recapitulating several
features of human disease (Johnson et al., 2008). In keeping with these results,
Tatom et al. (2009) have expressed human TDP-43 in the cytoplams of rat
neurons using an AAV vector and successfully reproduced several features of
FTD disease including gliosis, apoptosis, ubiquitination resulting in the eventual
development of motor impairments over time. However, as acknowledged by the
authors, the endogenous rat TDP-43 was still present in the injected neurons.
Therefore, it still remains to be established to what extent this situation occurs in
the neuronal human disease and to what extent these results truly mimic human
pathological mechanisms.

  In a rather distant species (but closer than yeast) such as *Drosophila*, it
has been shown that TDP-43 loss of function in the motoneurons leads to a
paralytic phenotype that can be rescued to a significant extent by pan-neuron
specific expression of both *Drosophila* and human TDP-43, further reinforcing
the conservation of the function and properties of TDP-43 through evolution
(Feiguin et al., 2009). These observations suggest that further effort should be
done to establish what proportion of the disease is due to toxicity of TDP-43
aggregates and what portion is due to loss of TDP-43 function in the nucleus.
These issues are discussed in Section IV.

  It was also logical to look for evidence of TDP-43 pathology in current
models of motor neuron diseases such as the SOD1 mice and the "wobbler"
model. Interestingly, two studies performed on different SOD1 mice genetic
lineages showed that there was no evidence of TDP-43 inclusions or mislocalization in the motor neurons in different lines of this type of mice (Robertson et al.,
2007; Turner et al., 2008). This result is in keeping with expectations based on
human patient's analyses, as lack of TDP-43 pathology in familial ALS cases
linked to SOD-1 mutations has been clearly highlighted in recent studies
(Mackenzie et al., 2007; Tan et al., 2007). This suggests that the mechanisms
of SOD-1 neurodegeneration may be functionally distinct from those mediated
by TDP-43. The work of Turner et al. (2008) also showed that no TDP-43
inclusions/mislocalization could be detected in SMN $-/-$ mice despite this
protein partially colocalized with the SMN protein itself, an observation that
is in keeping with previous data obtained using cell culture analyses (Wang et al.,
2002). This observation would tend to rule out a possible active role of TDP-43
in other type of motor neuron diseases such as Spinal Muscolar Atrophy although
the suggestion that TDP-43 may act as a modifier of SMN exon 7 splicing (Bose
et al., 2008) may change this picture in the future.

  In contrast, immunocytochemical studies performed on the "wobbler"
mice model revealed that TDP-43 was abnormally distributed in the cytoplasm
under the form of ubiquitinated inclusions and its expression levels were elevated

(2.2-fold increase) in the spinal cord of this type of mice (Dennis and Citron, 2009). The reason for this distribution may reside in the phenotypic defect of the "wobbler" mice that resides in the mutation of Vps54 (Schmitt-John et al., 2005), a Golgi-associated vacuolar protein sorting factor. Golgi disruption could thus be associated with defective clearance/transport of TDP-43 and thus in its aberrant localization and processing as previously described (Fujita et al., 2008). At the moment, therefore, the "wobbler" model seems to be the best TDP-43-related animal model of motor neuron disease.

IV. GAIN- VERSUS LOSS-OF-FUNCTION SCENARIOS

The first question that was asked with regards to the possible role played by TDP-43 in FTD/ALS was whether TDP-43 simply represented an indicator of disease or it was also an active player. The results obtained so far are all conducive to the conclusion that TDP-43 (or rather the lack of it, see below) represents an active player in neurodegeneration.

Regarding the TDP-43 role in the pathophysiological mechanisms of neurodegeneration, the issue that needs to be settled, briefly discussed above, is if the process is triggered by TDP-43 gain of function (toxic fragments), loss of function (absence from the nucleus), or a combination of both. In a loss-of-function scenario, nuclear depletion alone would be quite effective in causing a harmful effect on the cellular metabolism independently of the aberrant TDP-43 and/or the cytoplasmic aggregates having a potentially toxic effect. On the other hand, in a gain-of-function scenario the disruption of nuclear TDP-43 processes would be either very limited or reduced by compensatory mechanisms. In this situation, most harmful effects would be derived from direct toxicity of either the altered TDP-43 protein through phosphorylation/ubiquitination or the release of toxic C-terminal fragments following its degradation. These two scenarios are schematically summarized in Fig. 1.4 and Table 1.3.

For the moment, gain-of-function mechanisms are mainly supported by the observation that TDP-43 mutations act in a dominant form causing neurodegeneration in a gain-of-function context, either through increased toxicity of the degraded/modified protein or through the acquisition of novel biological properties. An example can be found in the previously discussed case of the D169G mutant that may have a better ability to withstand the action of ubiquilin-1 to segregate it in a cytoprotective aggregate. It should be noted though, that TDP-43 mutations are found in only 2–3% of ALS patients with TDP-43 aggregates and thus increased degradation/toxicity would account for increased pathogenicity only in a minority of patients. For this reason, mutations could then be considered more as a predisposing than a causative factor.

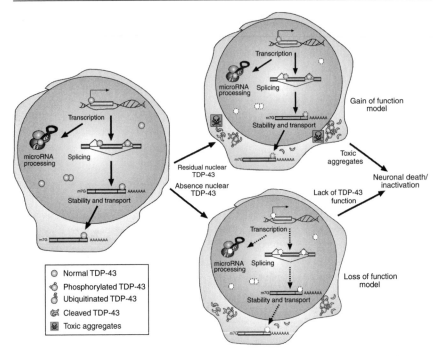

Figure 1.4. *Gain vs. Loss pathophysiological mechanisms.* Gain-of-function versus loss-of-function pathophysiological scenarios. In a gain-of-function situation, the toxic effects of the degraded fragments/aggregates are more harmful to the cell than the loss in nuclear TDP-43. In a loss-of-function situation, the cell suffers because all or critical TDP-43 controlled processes in the nucleus become mysregulated. The two situations are not mutually exclusive and a combination of both may actually contribute to cell death/inactivation.

On the other hand, TDP-43 mislocalization and degradation suggests that its absence from the nucleus may have dire consequences on the processes it controls, and thus pathogenicity would rather be the consequence of a loss-of-function scenario. In keeping with this view, TDP-43 knockout using siRNA techniques has been shown to result in misregulation of a myriad of cellular transcripts. This massive misregulation results in the observation that cells deprived of endogenous TDP-43 possessed a dysmorphic nuclear shape, misregulation of the cell cycle, and apoptosis (Ayala *et al.*, 2008a). Furthermore, it has been shown that aggregated TDP-43 mutants/variants are quite capable of sequestering endogenous TDP-43 within the insoluble aggregates (Gitcho *et al.*, 2008; Winton *et al.*, 2008a).

Therefore, the aggregates may not have a direct toxicity or, in a more extreme view, may even have a cytoprotective effect for the normal cellular metabolism. However, they might represent a "TDP-43 sink" that deprives the

Table 1.3. Evidence for TDP-43 Gain or Loss-of-Function Effects in Disease Development

Experimental results	Loss-of-function effects	Gain-of-function effects	References
siRNA-mediated TDP-43 knockout in culture cells	TDP-43 knockout results in dysmorphic nuclear shape, misregulation of the cell cycle, and apoptosis. In particular, protein and transcript levels of cyclin-dependent kinase 6 (Cdk6) and other factors that control cell proliferation significantly increase	–	Ayala *et al.* (2008a)
Drosophila melanogaster model	Flies lacking TBPH display a disease phenotype and show splicing defects for important genes involved in the neuromuscular junction formation and stress response	–	Feiguin *et al.* (2009)
Expression of C-terminal fragments in culture cells	Expression of a CTF fragment from residue 208 to 414 of TDP-43 can form cytoplasmic aggregates that are abnormally phosphorylated and ubiquitinated and its expression can alter the RNA splicing profiles of a CFTR exon 9 minigene	–	Igaz *et al.* (2009)
Expression of C-terminal fragments in yeast and culture cells	–	Expression of CTFs in yeast cells forms highly toxic and insoluble cytoplasmic inclusions	Johnson *et al.* (2008)

(*Continues*)

Table 1.3. (*Continued*)

Experimental results	Loss-of-function effects	Gain-of-function effects	References
Overexpression of wild-type human TDP-43 using AAVs in rat brains	–	Human TDP-43 overexpression in rat midbrain neurons results in gliosis, neuronal loss, and progressive motor dysfunctions in the injected animals. When expressed in the cytoplasm of midbrain neurons the TDP-43 distribution pattern was cytoplasmic and granular, consistent with preinclusion formation	Tatom *et al.* (2009)
Mutations in the TARDBP sequence	–	All mutations act in a dominant fashion. The presence of G38C, R361S, N390D/S mutations has been correlated with an increased presence of CTFs in lymphoblastoid preparations or cell lines. TDP-43 proteins carrying Q331K and M337V mutations have been associated with neurotoxicity when injected in chicken embryos	Kabashi *et al.* (2008) and Sreedharan *et al.* (2008)

CTFs, C-Terminal fragments.

cell not only of the aberrantly modified protein but also of the normal TDP-43. This situation will represent a loss-of-function scenario that in our opinion needs further consideration. The results obtained with the *Drosophila* model of TDP-43 depletion (see above) would be consistent with such hypothesis or with a combination of loss- and gain-of-function effects.

Indeed, this situation is quite reminiscent of the pathogenic mechanisms proposed for other neurodegenerative disease such as Alzheimer disease. If we take one of the players in this complex pathology, the presenilin gene, we can see that there is no clear winner between gain and loss-of-function models. In fact the mutations detected in the presenilin (*PSEN1/PSEN2*) genes are found associated with production of the Aβ42 product and with familial Alzheimer Disease.

As initial knockout studies of presenilins did not produce this phenotype the conclusion was that they acted through a gain-of-function mechanism. However, this conclusion has also been recently challenged by the observation that conditional *PSEN1/PSEN2* knockout mice displayed memory impairment and age-dependent neurodegeneration (Saura *et al.*, 2004) and that rare promoter mutations cause a significant drop in their transcriptional activity (Theuns *et al.*, 2000). Actually taken together, these data would rather suggest a loss-of-function role for presenilin genes in AD rather than the opposite.

For TDP-43, although the critical experiments that would definitively settle this question one way or the other are still lacking, the authors think that the existing experimental evidence suggests that TDP-43 proteinopathies may be more consistent with a loss-of-function scenario. Of course, it should always be kept in mind these two scenarios are not necessarily mutually exclusive and they may concur at the same (or different) time(s) to result in motor neuron death/inactivation.

Aside from basic science issues, establishing which scenario is better linked with neurodegeneration will be of fundamental importance to devise potential therapeutic strategies, which are schematically summarized in Fig. 1.5. Theoretically, in fact, a gain-of-function situation would be better addressed by strategies that target the mutated/modified molecule/fragment through the action of some inhibitor molecule. On the other hand, a loss-of-function situation would be better addressed by strategies that aim to keep functional TDP-43 above a critical threshold. This could be achieved either through the development of "effectors" capable of solubilizing/inhibiting aggregate formation or the exogenous introduction of TDP-43 in the cellular nucleus.

V. CONCLUSIONS

Despite recent advances, the mechanism(s) that link TDP-43 to this ever growing list of neurodegenerative pathologies still remain obscure. The discovery of disease-associated TDP-43 mutations and the observations that some TDP-43

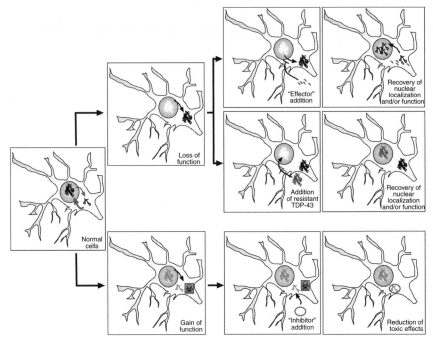

Figure 1.5. *Therapeutic strategies.* Potential therapeutic strategies dealing with eventual loss-of-function versus gain-of-function scenarios following the aberrant localization of TDP-43 in the cytoplasm and its degradation/modifications. The arrows in the diagram on the left indicate that TDP-43 normally shuttles between the nucleus to the cytoplasm although in normal conditions are much more abundant in the nuclear compartments. Neurons might be more sensitive than other types of cells to derangements in this situation due to their extended cytoplasms and specific needs on mRNA localization and transport state. In case of loss-of-function predominance, therapeutic recovery of this situation could be achieved either by the use of "effectors" (small molecules, etc.) which would interfere with aggregate formation or solubilize existing ones to release wild-type TDP-43. An alternative would be to provide the cell with an aggregate-resistant form of TDP-43. On the other hand, in the case of predominance by gain-of-function effects it might be more profitable to devise inhibitors capable of reducing the agents responsible for this toxicity (i.e., C-terminal fragments, etc.).

modifications (degradation, aggregation, phosphorylation) closely correlate with disease tell us that a link does indeed exist. At present, the greatest impact of TDP-43 research has been on the clinical and diagnostic community. As far as the clinic is concerned, there has been a complete reappraisal of old classifications which will surely be beneficial to prognostic and patient management protocols (Mackenzie *et al.*, 2009). In addition, the diagnostic field is currently searching for novel TDP-43 markers that might provide easier screening procedures to identify these diseases at an earlier stage than currently possible (Foulds *et al.*, 2008; Kasai *et al.*, 2009; Steinacker *et al.*, 2008).

Nonetheless, there are questions for which there is still no answer: for example, how and why TDP-43 modifications occur (sequentially?, in parallel?, independently?) and what is their pathological potential? Is TDP-43 loss-of-function or gain-of-function critical in ALS pathogenesis? Which are the target mRNAs whose function in motoneurons will be altered by TDP-43 unavailability? These are not just scientific curiosities because, as discussed in the sections above, knowing the answers to at least some of these questions will profoundly influence the future search for potential therapeutic strategies. An additional element that is still sorely lacking in TDP-43 research is the availability of TDP-43-based animal models. In this review we have reported the current knowledge with regards to established animal models of motor neuron disease (SOD-1 and Wobbler mice) but these will only be able to provide limited information with regards to TDP-43 pathology. This should be a gap, however, that might be filled very soon if we consider the attention that TDP-43 has been given by the neurodegenerative disease research community.

Acknowledgments

This work was supported by Telethon Onlus Foundation (Italy) (grant no. GGP06147) and by a European community grant (EURASNET-LSHG-CT-2005-518238).

References

Abhyankar, M. M., Urekar, C., and Reddi, P. P. (2007). A novel CpG-free vertebrate insulator silences the testis-specific SP-10 gene in somatic tissues: Role for TDP-43 in insulator function. *J. Biol. Chem.* **282**, 36143–36154.

Acharya, K. K., Govind, C. K., Shore, A. N., Stoler, M. H., and Reddi, P. P. (2006). cis-Requirement for the maintenance of round spermatid-specific transcription. *Dev. Biol.* **295**, 781–790.

Ahmed, Z., Mackenzie, I. R., Hutton, M. L., and Dickson, D. W. (2007). Progranulin in frontotemporal lobar degeneration and neuroinflammation. *J. Neuroinflammation* **4**, 7.

Amador-Ortiz, C., Lin, W. L., Ahmed, Z., Personett, D., Davies, P., Duara, R., Graff-Radford, N. R., Hutton, M. L., and Dickson, D. W. (2007). TDP-43 immunoreactivity in hippocampal sclerosis and Alzheimer's disease. *Ann. Neurol.* **61**, 435–445.

Arai, T., Hasegawa, M., Akiyama, H., Ikeda, K., Nonaka, T., Mori, H., Mann, D., Tsuchiya, K., Yoshida, M., Hashizume, Y., and Oda, T. (2006). TDP-43 is a component of ubiquitin-positive tau-negative inclusions in frontotemporal lobar degeneration and amyotrophic lateral sclerosis. *Biochem. Biophys. Res. Commun.* **351**, 602–611.

Arai, T., Mackenzie, I. R., Hasegawa, M., Nonoka, T., Niizato, K., Tsuchiya, K., Iritani, S., Onaya, M., and Akiyama, H. (2009). Phosphorylated TDP-43 in Alzheimer's disease and dementia with Lewy bodies. *Acta Neuropathol.* **117**, 125–136.

Arrisi-Mercado, P., Romano, M., Muro, A. F., and Baralle, F. E. (2004). An exonic splicing enhancer offsets the atypical GU-rich 3′ splice site of human apolipoprotein A-II exon 3. *J. Biol. Chem.* **279,** 39331–39339.

Ayala, Y. M., Pantano, S., D'Ambrogio, A., Buratti, E., Brindisi, A., Marchetti, C., Romano, M., and Baralle, F. E. (2005). Human, Drosophila, and C. elegans TDP43: Nucleic acid binding properties and splicing regulatory function. *J. Mol. Biol.* **348,** 575–588.

Ayala, Y. M., Misteli, T., and Baralle, F. E. (2008a). TDP-43 regulates retinoblastoma protein phosphorylation through the repression of cyclin-dependent kinase 6 expression. *Proc. Natl. Acad. Sci. USA* **105,** 3785–3789.

Ayala, Y. M., Zago, P., D'Ambrogio, A., Xu, Y. F., Petrucelli, L., Buratti, E., and Baralle, F. E. (2008b). Structural determinants of the cellular localization and shuttling of TDP-43. *J. Cell Sci.* **121,** 3778–3785.

Banks, G. T., Kuta, A., Isaacs, A. M., and Fisher, E. M. (2008). TDP-43 is a culprit in human neurodegeneration, and not just an innocent bystander. *Mamm. Genome* **19,** 299–305.

Bigio, E. H. (2008). TAR DNA-binding protein-43 in amyotrophic lateral sclerosis, frontotemporal lobar degeneration, and Alzheimer disease. *Acta Neuropathol.* **116,** 135–140.

Bose, J. K., Wang, I. F., Hung, L., Tarn, W. Y., and Shen, C. K. (2008). TDP-43 overexpression enhances exon 7 inclusion during the survival of motor neuron pre-mRNA splicing. *J. Biol. Chem.* **283,** 28852–28859.

Bugiani, O. (2007). The many ways to frontotemporal degeneration and beyond. *Neurol. Sci.* **28,** 241–244.

Buratti, E., and Baralle, F. E. (2001). Characterization and functional implications of the RNA binding properties of nuclear factor TDP-43, a novel splicing regulator of CFTR exon 9. *J. Biol. Chem.* **276,** 36337–36343.

Buratti, E., and Baralle, F. E. (2008). Multiple roles of TDP-43 in gene expression, splicing regulation, and human disease. *Front Biosci.* **13,** 867–878.

Buratti, E., Dork, T., Zuccato, E., Pagani, F., Romano, M., and Baralle, F. E. (2001). Nuclear factor TDP-43 and SR proteins promote *in vitro* and *in vivo* CFTR exon 9 skipping. *EMBO J.* **20,** 1774–1784.

Buratti, E., Brindisi, A., Giombi, M., Tisminetzky, S., Ayala, Y. M., and Baralle, F. E. (2005). TDP-43 Binds Heterogeneous Nuclear Ribonucleoprotein A/B through Its C-terminal Tail: an important region for the inhibition of cystic fibrosis transmembrane conductance regulator exon 9 splicing. *J. Biol. Chem.* **280,** 37572–37584.

Carpenter, B., MacKay, C., Alnabulsi, A., MacKay, M., Telfer, C., Melvin, W. T., and Murray, G. I. (2006). The roles of heterogeneous nuclear ribonucleoproteins in tumour development and progression. *Biochim. Biophys. Acta* **1765,** 85–100.

Cook, C., Zhang, Y. J., Xu, Y. F., Dickson, D. W., and Petrucelli, L. (2008). TDP-43 in neurodegenerative disorders. *Expert Opin. Biol. Ther.* **8,** 969–978.

Corrado, L., Ratti, A., Gellera, C., Buratti, E., Castellotti, B., Carlomagno, Y., Ticozzi, N., Mazzini, L., Testa, L., Taroni, F., Baralle, F. E., Silani, V., *et al.* (2009). High frequency of TARDBP gene mutations in Italian patients with amyotrophic lateral sclerosis. *Hum. Mutat.* **30,** 688–694.

Daoud, H., Valdmanis, P. N., Kabashi, E., Dion, P., Dupre, N., Camu, W., Meininger, V., and Rouleau, G. A. (2008). Contribution of TARDBP mutations to sporadic amyotrophic lateral sclerosis. *J. Med. Genet.* **46,** 112–114.

Del Bo, R., Ghezzi, S., Corti, S., Pandolfo, M., Ranieri, M., Santoro, D., Ghione, I., Prelle, A., Orsetti, V., Mancuso, M., Sorarù, G., Briani, C., *et al.* (2009). TARDBP (TDP-43) sequence analysis in patients with familial and sporadic ALS: Identification of two novel mutations. *Eur. J. Neurol.* **16,** 727–732.

Dennis, J. S., and Citron, B. A. (2009). Wobbler mice modeling motor neuron disease display elevated transactive response DNA binding protein. *Neuroscience* **158**, 745–750.

Dickson, D. W. (2008). TDP-43 immunoreactivity in neurodegenerative disorders: disease versus mechanism specificity. *Acta Neuropathol.* **115**, 147–149.

Dickson, D. W., Josephs, K. A., and Amador-Ortiz, C. (2007). TDP-43 in differential diagnosis of motor neuron disorders. *Acta Neuropathol. (Berl)* **114**, 71–79.

Dreyfuss, G., Kim, V. N., and Kataoka, N. (2002). Messenger-RNA-binding proteins and the messages they carry. *Nat. Rev. Mol. Cell Biol.* **3**, 195–205.

Elman, L. B., McCluskey, L., and Grossman, M. (2008). Motor neuron disease and frontotemporal lobar degeneration: A tale of two disorders linked to TDP-43. *Neurosignals* **16**, 85–90.

Elvira, G., Wasiak, S., Blandford, V., Tong, X. K., Serrano, A., Fan, X., del Rayo Sanchez-Carbente, M., Servant, F., Bell, A. W., Boismenu, D., Lacaille, J. C., McPherson, P. S., *et al.* (2006). Characterization of an RNA granule from developing brain. *Mol. Cell Proteomics* **5**, 635–651.

Farrer, M. J., Hulihan, M. M., Kachergus, J. M., Dachsel, J. C., Stoessl, A. J., Grantier, L. L., Calne, S., Calne, D. B., Lechevalier, B., Chapon, F., Tsuboi, Y., Yamada, T., *et al.* (2009). DCTN1 mutations in Perry syndrome. *Nat. Genet.* **41**, 163–165.

Feiguin, F., Godena, V. K., Romano, G., D'Ambrogio, A., Klima, R., and Baralle, F. E. (2009). Depletion of TDP-43 affects Drosophila motoneurons terminal synapsis and locomotive behaviour. *FEBS Lett.* **583**, 1586–1592.

Filimonenko, M., Stuffers, S., Raiborg, C., Yamamoto, A., Malerod, L., Fisher, E. M., Isaacs, A., Brech, A., Stenmark, H., and Simonsen, A. (2007). Functional multivesicular bodies are required for autophagic clearance of protein aggregates associated with neurodegenerative disease. *J. Cell Biol.* **179**, 485–500.

Forman, M. S., Trojanowski, J. Q., and Lee, V. M. (2007). TDP-43: A novel neurodegenerative proteinopathy. *Curr. Opin. Neurobiol.* **17**, 548–555.

Foulds, P., McAuley, E., Gibbons, L., Davidson, Y., Pickering-Brown, S. M., Neary, D., Snowden, J. S., Allsop, D., and Mann, D. M. (2008). TDP-43 protein in plasma may index TDP-43 brain pathology in Alzheimer's disease and frontotemporal lobar degeneration. *Acta Neuropathol.* **116**, 141–146.

Fujita, Y., Mizuno, Y., Takatama, M., and Okamoto, K. (2008). Anterior horn cells with abnormal TDP-43 immunoreactivities show fragmentation of the Golgi apparatus in ALS. *J. Neurol. Sci.* **269**, 30–34.

Fukuda, T., Yamagata, K., Fujiyama, S., Matsumoto, T., Koshida, I., Yoshimura, K., Mihara, M., Naitou, M., Endoh, H., Nakamura, T., Akimoto, C., Yamamoto, Y., *et al.* (2007). DEAD-box RNA helicase subunits of the Drosha complex are required for processing of rRNA and a subset of microRNAs. *Nat. Cell Biol.* **9**, 604–611.

Geser, F., Martinez-Lage, M., Kwong, L. K., Lee, V. M., and Trojanowski, J. Q. (2009a). Amyotrophic lateral sclerosis, frontotemporal dementia and beyond: The TDP-43 diseases. *J. Neurol.* In press.

Geser, F., Martinez-Lage, M., Robinson, J., Uryu, K., Neumann, M., Brandmeir, N. J., Xie, S. X., Kwong, L. K., Elman, L., McCluskey, L., Clark, C. M., Malunda, J., *et al.* (2009b). Clinical and pathological continuum of multisystem TDP-43 proteinopathies. *Arch. Neurol.* **66**, 180–189.

Gijselinck, I., Sleegers, K., Engelborghs, S., Robberecht, W., Martin, J. J., Vandenberghe, R., Sciot, R., Dermaut, B., Goossens, D., van der Zee, J., De Pooter, T., Del Favero, J., *et al.* (2007). Neuronal inclusion protein TDP-43 has no primary genetic role in FTD and ALS. *Neurobiol. Aging.* **30**, 1329–1331.

Gitcho, M. A., Baloh, R. H., Chakraverty, S., Mayo, K., Norton, J. B., Levitch, D., Hatanpaa, K. J., White, C. L., 3rd, Bigio, E. H., Caselli, R., Baker, M., Al-Lozi, M. T., *et al.* (2008). TDP-43 A315T mutation in familial motor neuron disease. *Ann. Neurol.* **63**, 535–538.

Gitcho, M. A., Strider, J., Carter, D., Taylor-Reinwald, L., Forman, M. S., Goate, A. M., and Cairns, N. J. (2009). VCP mutations causing frontotemporal lobar degeneration disrupt localization of TDP-43 and induce cell death. *J. Biol. Chem.* **284,** 12384–12398.

Glisovic, T., Bachorik, J. L., Yong, J., and Dreyfuss, G. (2008). RNA-binding proteins and post-transcriptional gene regulation. *FEBS Lett.* **582,** 1977–1986.

Gregory, R. I., Yan, K. P., Amuthan, G., Chendrimada, T., Doratotaj, B., Cooch, N., and Shiekhattar, R. (2004). The Microprocessor complex mediates the genesis of microRNAs. *Nature* **432,** 235–240.

Guerreiro, R. J., Schymick, J. C., Crews, C., Singleton, A., Hardy, J., and Traynor, B. J. (2008). TDP-43 is not a common cause of sporadic amyotrophic lateral sclerosis. *PLoS ONE* **3,** e2450.

Hasegawa, M., Arai, T., Nonaka, T., Kametani, F., Yoshida, M., Hashizume, Y., Beach, T. G., Buratti, E., Baralle, F., Morita, M., *et al.* (2008). Phosphorylated TDP-43 in frontotemporal lobar degeneration and amyotrophic lateral sclerosis. *Ann. Neurol.* **64,** 60–70.

He, Y., and Smith, R. (2007). Nuclear functions of heterogeneous nuclear ribonucleoproteins A/B. *Cell Mol. Life Sci.* **114,** 71–79.

Igaz, L. M., Kwong, L. K., Xu, Y., Truax, A. C., Uryu, K., Neumann, M., Clark, C. M., Elman, L. B., Miller, B. L., Grossman, M., McCluskey, L. F., Trojanowski, J. Q., *et al.* (2008). Enrichment of C-terminal fragments in TAR DNA-binding protein-43 cytoplasmic inclusions in brain but not in spinal cord of frontotemporal lobar degeneration and amyotrophic lateral sclerosis. *Am. J. Pathol.* **173,** 182–194.

Igaz, L. M., Kwong, L. K., Chen-Plotkin, A., Winton, M. J., Unger, T. L., Xu, Y., Neumann, M., Trojanowski, J. Q., and Lee, V. M. (2009). Expression of TDP-43 C-terminal fragments *in vitro* recapitulates pathological features of TDP-43 proteinopathies. *J. Biol. Chem.* **284,** 8516–8524.

Inukai, Y., Nonaka, T., Arai, T., Yoshida, M., Hashizume, Y., Beach, T. G., Buratti, E., Baralle, F. E., Akiyama, H., Hisanaga, S., and Hasegawa, M. (2008). Abnormal phosphorylation of Ser409/410 of TDP-43 in FTLD-U and ALS. *FEBS Lett.* **582,** 2899–2904.

Johnson, B. S., McCaffery, J. M., Lindquist, S., and Gitler, A. D. (2008). A yeast TDP-43 proteinopathy model: Exploring the molecular determinants of TDP-43 aggregation and cellular toxicity. *Proc. Natl. Acad. Sci. USA* **105,** 6439–6444.

Kabashi, E., Valdmanis, P. N., Dion, P., Spiegelman, D., McConkey, B. J., Vande Velde, C., Bouchard, J. P., Lacomblez, L., Pochigaeva, K., Salachas, F., Pradat, P. F., Camu, W., *et al.* (2008). TARDBP mutations in individuals with sporadic and familial amyotrophic lateral sclerosis. *Nat. Genet.* **40,** 572–574.

Kabashi, E., Daoud, H., Riviere, J. B., Valdamanis, P. N., Bourgouin, P., Provencher, P., Pourcher, E., Dion, P., Dupre, N., and Rouleau, G. A. (2009). No TARDBP mutations in a French Canadian population of patients with Parkinson disease. *Arch. Neurol.* **66,** 281–282.

Kasai, T., Tokuda, T., Ishigami, N., Sasayama, H., Foulds, P., Mitchell, D. J., Mann, D. M., Allsop, D., and Nakagawa, M. (2009). Increased TDP-43 protein in cerebrospinal fluid of patients with amyotrophic lateral sclerosis. *Acta Neuropathol.* **117,** 55–62.

Kato, S. (2008). Amyotrophic lateral sclerosis models and human neuropathology: similarities and differences. *Acta Neuropathol.* **115,** 97–114.

Kim, S. H., Shi, Y., Hanson, K. A., Williams, L. M., Sakasai, R., Bowler, M. J., and Tibbetts, R. S. (2008). Potentiation of ALS-associated TDP-43 aggregation by the proteasome-targeting factor, Ubiquilin 1. *J. Biol. Chem.* **284,** 8083–8092.

Krecic, A. M., and Swanson, M. S. (1999). hnRNP complexes: Composition, structure, and function. *Curr. Opin. Cell Biol.* **11,** 363–371.

Kuhnlein, P., Sperfeld, A. D., Vanmassenhove, B., Van Deerlin, V., Lee, V. M., Trojanowski, J. Q., Kretzschmar, H. A., Ludolph, A. C., and Neumann, M. (2008). Two German kindreds with familial amyotrophic lateral sclerosis due to TARDBP mutations. *Arch. Neurol.* **65,** 1185–1189.

Kuo, P. H., Doudeva, L. G., Wang, Y. T., Shen, C. K., and Yuan, H. S. (2009). Structural insights into TDP-43 in nucleic-acid binding and domain interactions. *Nucleic Acids. Res.* **37,** 1799–1808.

Kwiatkowski, T. J., Jr., Bosco, D. A., Leclerc, A. L., Tamrazian, E., Vanderburg, C. R., Russ, C., Davis, A., Gilchrist, J., Kasarskis, E. J., Munsat, T., Valdmanis, P., Rouleau, G. A., *et al.* (2009). Mutations in the FUS/TLS gene on chromosome 16 cause familial amyotrophic lateral sclerosis. *Science* **323,** 1205–1208.

Kwong, L. K., Uryu, K., Trojanowski, J. Q., and Lee, V. M. (2008). TDP-43 proteinopathies: Neurodegenerative protein misfolding diseases without amyloidosis. *Neurosignals* **16,** 41–51.

Lee, E. B., Lee, V. M., Trojanowski, J. Q., and Neumann, M. (2008). TDP-43 immunoreactivity in anoxic, ischemic and neoplastic lesions of the central nervous system. *Acta Neuropathol.* **115,** 305–311.

Lehner, B., and Sanderson, C. M. (2004). A protein interaction framework for human mRNA degradation. *Genome Res.* **14,** 1315–1323.

Lemmens, R., Race, V., Hersmus, N., Matthijs, G., Van Den Bosch, L., Van Damme, P., Dubois, B., Boonen, S., Goris, A., and Robberecht, W. (2009). TDP-43 M311V mutation in familial amyotrophic lateral sclerosis. *J. Neurol. Neurosurg. Psychiatry* **80,** 354–355.

Liscic, R. M., Grinberg, L. T., Zidar, J., Gitcho, M. A., and Cairns, N. J. (2008). ALS and FTLD: Two faces of TDP-43 proteinopathy. *Eur. J. Neurol.* **15,** 772–780.

Mackenzie, I. R., and Rademakers, R. (2007). The molecular genetics and neuropathology of frontotemporal lobar degeneration: Recent developments. *Neurogenetics* **8,** 237–248.

Mackenzie, I. R., and Rademakers, R. (2008). The role of transactive response DNA-binding protein-43 in amyotrophic lateral sclerosis and frontotemporal dementia. *Curr. Opin. Neurol.* **21,** 693–700.

Mackenzie, I. R., Bigio, E. H., Ince, P. G., Geser, F., Neumann, M., Cairns, N. J., Kwong, L. K., Forman, M. S., Ravits, J., Stewart, H., Eisen, A., McClusky, L., *et al.* (2007). Pathological TDP-43 distinguishes sporadic amyotrophic lateral sclerosis from amyotrophic lateral sclerosis with SOD1 mutations. *Ann. Neurol.* **61,** 427–434.

Mackenzie, I. R., Neumann, M., Bigio, E. H., Cairns, N. J., Alafuzoff, I., Kril, J., Kovacs, G. G., Ghetti, B., Halliday, G., Holm, I. E., Ince, P. G., Kamphorst, W., *et al.* (2009). Nomenclature for neuropathologic subtypes of frontotemporal lobar degeneration: Consensus recommendations. *Acta Neuropathol.* **117,** 15–18.

Martinez-Contreras, R., Cloutier, P., Shkreta, L., Fisette, J. F., Revil, T., and Chabot, B. (2007). hnRNP proteins and splicing control. *Adv. Exp. Med. Biol.* **623,** 123–147.

Mercado, P. A., Ayala, Y. M., Romano, M., Buratti, E., and Baralle, F. E. (2005). Depletion of TDP-43 overrides the need for exonic and intronic splicing enhancers in the human apoA-II gene. *Nucleic Acids Res.* **33,** 6000–6010.

Moisse, K., Volkening, K., Leystra-Lantz, C., Welch, I., Hill, T., and Strong, M. J. (2009). Divergent patterns of cytosolic TDP-43 and neuronal progranulin expression following axotomy: Implications for TDP-43 in the physiological response to neuronal injury. *Brain Res.* **1249,** 202–211.

Nakamura, M., Ito, H., Wate, R., Nakano, S., Hirano, A., and Kusaka, H. (2008). Phosphorylated Smad2/3 immunoreactivity in sporadic and familial amyotrophic lateral sclerosis and its mouse model. *Acta Neuropathol.* **115,** 327–334.

Neumann, M., Igaz, L. M., Kwong, L. K., Nakashima-Yasuda, H., Kolb, S. J., Dreyfuss, G., Kretzschmar, H. A., Trojanowski, J. Q., and Lee, V. M. (2007a). Absence of heterogeneous nuclear ribonucleoproteins and survival motor neuron protein in TDP-43 positive inclusions in frontotemporal lobar degeneration. *Acta Neuropathol. (Berl)* **113,** 543–548.

Neumann, M., Kwong, L. K., Sampathu, D. M., Trojanowski, J. Q., and Lee, V. M. (2007b). TDP-43 proteinopathy in frontotemporal lobar degeneration and amyotrophic lateral sclerosis: protein misfolding diseases without amyloidosis. *Arch. Neurol.* **64,** 1388–1394.

Neumann, M., Sampathu, D. M., Kwong, L. K., Truax, A. C., Micsenyi, M. C., Chou, T. T., Bruce, J., Schuck, T., Grossman, M., Clark, C. M., McCluskey, L. F., Miller, B. L., et al. (2006). Ubiquitinated TDP-43 in frontotemporal lobar degeneration and amyotrophic lateral sclerosis. Science 314, 130–133.

Neumann, M., Kwong, L. K., Lee, E. B., Kremmer, E., Flatley, A., Xu, Y., Forman, M. S., Troost, D., Kretzschmar, H. A., Trojanowski, J. Q., and Lee, V. M. (2009). Phosphorylation of S409/410 of TDP-43 is a consistent feature in all sporadic and familial forms of TDP-43 proteinopathies. Acta Neuropathol. 117, 137–149.

Nonaka, T., Arai, T., Buratti, E., Baralle, F. E., Akiyama, H., and Hasegawa, M. (2008). Phosphorylated and ubiquitinated TDP-43 pathological inclusions in ALS and FTLD-U are recapitulated in SH-SY5Y cells. FEBS Lett. 583, 394–400.

Ou, S. H., Wu, F., Harrich, D., Garcia-Martinez, L. F., and Gaynor, R. B. (1995). Cloning and characterization of a novel cellular protein, TDP-43, that binds to human immunodeficiency virus type 1 TAR DNA sequence motifs. J. Virol. 69, 3584–3596.

Pickering-Brown, S. M. (2007). Progranulin and frontotemporal lobar degeneration. Acta Neuropathol. 114, 39–47.

Robertson, J., Sanelli, T., Xiao, S., Yang, W., Horne, P., Hammond, R., Pioro, E. P., and Strong, M. J. (2007). Lack of TDP-43 abnormalities in mutant SOD1 transgenic mice shows disparity with ALS. Neurosci. Lett. 420, 128–132.

Rohn, T. T. (2008). Caspase-cleaved TAR DNA-binding protein-43 is a major pathological finding in Alzheimer's disease. Brain Res. 1228, 189–198.

Rollinson, S., Snowden, J. S., Neary, D., Morrison, K. E., Mann, D. M., and Pickering-Brown, S. M. (2007). TDP-43 gene analysis in frontotemporal lobar degeneration. Neurosci. Lett. 419, 1–4.

Rollinson, S., Rizzu, P., Sikkink, S., Baker, M., Halliwell, N., Snowden, J., Traynor, B. J., Ruano, D., Cairns, N., Rohrer, J. D., Mead, S., Collinge, J., et al. (2009). Ubiquitin associated protein 1 is a risk factor for frontotemporal lobar degeneration. Neurobiol. Aging. 30, 656–665.

Rothstein, J. D. (2007). TDP-43 in amyotrophic lateral sclerosis: Pathophysiology or patho-babel? Ann. Neurol. 61, 382–384.

Rutherford, N. J., Zhang, Y. J., Baker, M., Gass, J. M., Finch, N. A., Xu, Y. F., Stewart, H., Kelley, B. J., Kuntz, K., Crook, R. J., Sreedharan, J., Vance, C., et al. (2008). Novel mutations in TARDBP (TDP-43) in patients with familial amyotrophic lateral sclerosis. PLoS Genet. 4, e1000193.

Saura, C. A., Choi, S. Y., Beglopoulos, V., Malkani, S., Zhang, D., Shankaranarayana Rao, B. S., Chattarji, S., Kelleher, R. J., 3rd, Kandel, E. R., Duff, K., Kirkwood, A., and Shen, J. (2004). Loss of presenilin function causes impairments of memory and synaptic plasticity followed by age-dependent neurodegeneration. Neuron 42, 23–36.

Schmitt-John, T., Drepper, C., Mussmann, A., Hahn, P., Kuhlmann, M., Thiel, C., Hafner, M., Lengeling, A., Heimann, P., Jones, J. M., Meisler, M. H., and Jockusch, H. (2005). Mutation of Vps54 causes motor neuron disease and defective spermiogenesis in the wobbler mouse. Nat. Genet. 37, 1213–1215.

Schumacher, A., Friedrich, P., Diehl-Schmid, J., Ibach, B., Perneczky, R., Eisele, T., Vukovich, R., Foerstl, H., and Riemenschneider, M. (2009). No association of TDP-43 with sporadic frontotemporal dementia. Neurobiol. Aging. 30, 157–159.

Segref, A., and Hoppe, T. (2009). Think locally: control of ubiquitin-dependent protein degradation in neurons. EMBO Rep. 10, 44–50.

Shankaran, S. S., Capell, A., Hruscha, A. T., Fellerer, K., Neumann, M., Schmid, B., and Haass, C. (2008). Missense mutations in the progranulin gene linked to frontotemporal lobar degeneration with ubiquitin-immunoreactive inclusions reduce progranulin production and secretion. J. Biol. Chem. 283, 1744–1753.

Sorarù, G., Orsetti, V., Buratti, E., Baralle, F. E., Cima, V., Volpe, M., D'Ascenzo, C., Palmieri, A., Koutsikos, K., Pegoraro, E., and Angelini, C. (2009). TDP-43 in skeletal muscle of patients affected with amyotrophic lateral sclerosis. *Amyotroph. Lateral. Scler.* **20**, 1–5.

Sreedharan, J., Blair, I. P., Tripathi, V. B., Hu, X., Vance, C., Rogelj, B., Ackerley, S., Durnall, J. C., Williams, K. L., Buratti, E., Baralle, F., de Belleroche, J., *et al.* (2008). TDP-43 mutations in familial and sporadic amyotrophic lateral sclerosis. *Science* **319**, 1668–1672.

Steinacker, P., Hendrich, C., Sperfeld, A. D., Jesse, S., von Arnim, C. A., Lehnert, S., Pabst, A., Uttner, I., Tumani, H., Lee, V. M., Trojanowski, J. Q., Kretzschmar, H. A., *et al.* (2008). TDP-43 in cerebrospinal fluid of patients with frontotemporal lobar degeneration and amyotrophic lateral sclerosis. *Arch. Neurol.* **65**, 1481–1487.

Stelzl, U., Worm, U., Lalowski, M., Haenig, C., Brembeck, F. H., Goehler, H., Stroedicke, M., Zenkner, M., Schoenherr, A., Koeppen, S., Timm, J., Mintzlaff, S., *et al.* (2005). A human protein-protein interaction network: A resource for annotating the proteome. *Cell* **122**, 957–968.

Strong, M. J., Volkening, K., Hammond, R., Yang, W., Strong, W., Leystra-Lantz, C., and Shoesmith, C. (2007). TDP43 is a human low molecular weight neurofilament (hNFL) mRNA-binding protein. *Mol. Cell Neurosci.* **35**, 320–327.

Tan, C. F., Eguchi, H., Tagawa, A., Onodera, O., Iwasaki, T., Tsujino, A., Nishizawa, M., Kakita, A., and Takahashi, H. (2007). TDP-43 immunoreactivity in neuronal inclusions in familial amyotrophic lateral sclerosis with or without SOD1 gene mutation. *Acta Neuropathol. (Berl)* **113**, 535–542.

Tatom, J. B., Wang, D. B., Dayton, R. D., Skalli, O., Hutton, M. L., Dickson, D. W., and Klein, R. L. (2009). Mimicking aspects of frontotemporal lobar degeneration and Lou Gehrig's disease in rats via TDP-43 overexpression. *Mol. Ther.* **17**, 607–613.

Theuns, J., Del-Favero, J., Dermaut, B., van Duijn, C. M., Backhovens, H., Van den Broeck, M. V., Serneels, S., Corsmit, E., Van Broeckhoven, C. V., and Cruts, M. (2000). Genetic variability in the regulatory region of presenilin 1 associated with risk for Alzheimer's disease and variable expression. *Hum. Mol. Genet.* **9**, 325–331.

Tolnay, M., and Frank, S. (2007). Pathology and genetics of frontotemporal lobar degeneration: An update. *Clin. Neuropathol.* **26**, 143–156.

Turner, B. J., Baumer, D., Parkinson, N. J., Scaber, J., Ansorge, O., and Talbot, K. (2008). TDP-43 expression in mouse models of amyotrophic lateral sclerosis and spinal muscular atrophy. *BMC Neurosci.* **9**, 104.

Vance, C., Rogelj, B., Hortobagyi, T., De Vos, K. J., Nishimura, A. L., Sreedharan, J., Hu, X., Smith, B., Ruddy, D., Wright, P., Ganesalingam, J., Williams, K. L., *et al.* (2009). Mutations in FUS, an RNA processing protein, cause familial amyotrophic lateral sclerosis type 6. *Science* **323**, 1208–1211.

Van Deerlin, V. M., Leverenz, J. B., Bekris, L. M., Bird, T. D., Yuan, W., Elman, L. B., Clay, D., Wood, E. M., Chen-Plotkin, A. S., Martinez-Lage, M., Steinbart, E., McCluskey, L., *et al.* (2008). TARDBP mutations in amyotrophic lateral sclerosis with TDP-43 neuropathology: A genetic and histopathological analysis. *Lancet Neurol.* **7**, 409–416.

Wang, I. F., Reddy, N. M., and Shen, C. K. (2002). Higher order arrangement of the eukaryotic nuclear bodies. *Proc. Natl. Acad. Sci. USA* **99**, 13583–13588.

Wang, I. F., Wu, L. S., Chang, H. Y., and Shen, C. K. (2008a). TDP-43, the signature protein of FTLD-U, is a neuronal activity-responsive factor. *J. Neurochem.* **105**, 797–806.

Wang, I. F., Wu, L. S., and Shen, C. K. (2008b). TDP-43: An emerging new player in neurodegenerative diseases. *Trends Mol. Med.* **14**, 479–485.

Wider, C., Dickson, D. W., Stoessl, A. J., Tsuboi, Y., Chapon, F., Gutmann, L., Lechevalier, B., Calne, D. B., Personett, D. A., Hulihan, M., Kachergus, J., Rademakers, R., *et al.* (2008). Pallidonigral TDP-43 pathology in Perry syndrome. *Parkinsonism Relat. Disord.* **15**, 281–286.

Winton, M. J., Igaz, L. M., Wong, M. M., Kwong, L. K., Trojanowski, J. Q., Lee, V. M., Trojanowski, J. Q., and Miller, B. L. (2008a). Disturbance of nuclear and cytoplasmic TAR DNA-binding protein (TDP-43) induces disease-like redistribution, sequestration, and aggregate formation. *J. Biol. Chem.* **283**, 13302–13309.

Winton, M. J., Van Deerlin, V. M., Kwong, L. K., Yuan, W., Wood, E. M., Yu, C. E., Schellenberg, G. D., Rademakers, R., Caselli, R., Karydas, A., Trojanowski, J. Q., Miller, B. L., *et al.* (2008b). A90V TDP-43 variant results in the aberrant localization of TDP-43 *in vitro*. *FEBS Lett.* **582**, 2252–2256.

Yokoseki, A., Shiga, A., Tan, C. F., Tagawa, A., Kaneko, H., Koyama, A., Eguchi, H., Tsujino, A., Ikeuchi, T., Kakita, A., Okamoto, K., Nishizawa, M., *et al.* (2008). TDP-43 mutation in familial amyotrophic lateral sclerosis. *Ann. Neurol.* **63**, 538–542.

Zhang, Y. J., Xu, Y. F., Dickey, C. A., Buratti, E., Baralle, F., Bailey, R., Pickering-Brown, S., Dickson, D., and Petrucelli, L. (2007). Progranulin mediates caspase-dependent cleavage of TAR DNA binding protein-43. *J. Neurosci.* **27**, 10530–10534.

Zhang, H. X., Tanji, K., Mori, F., and Wakabayashi, K. (2008). Epitope mapping of 2E2-D3, a monoclonal antibody directed against human TDP-43. *Neurosci. Lett.* **434**, 170–174.

2

Resources and Strategies to Integrate the Study of Ethical, Legal, and Social Implications of Genetics into the Undergraduate Curriculum

Jinnie M. Garrett* and Kathleen L. Triman†

*Department of Biology, Hamilton College, Clinton, New York, NY, USA
†Department of Biology, Franklin and Marshall College, Lancaster, PA, USA

ABSTRACT

Gene therapy, genetically modified organisms, and the privacy of an individual's genetic information are just a few of the developments emerging from recent advances in molecular genetics that are controversial. Oversight and regulation

Advances in Genetics, Vol. 66
Copyright 2009, Elsevier Inc. All rights reserved.

of emerging technologies are the responsibility of both experts and the general public who both need to understand the science and the societal impact of its use. The study of ethical, legal, and social implications (ELSI) of advances in genetics provides a very powerful pedagogical tool to accomplish two goals. These are, first of all, to interest nonscientists in genetics and engage them in learning the science behind the ELSI developments they are considering, and secondly, to broaden the perspective of science students to consider the history and social consequences of the science they are studying. The resources and strategies presented in this chapter for teaching ELSI issues that arise in modern genetics are designed to aid in accomplishing these goals throughout the under-graduate curriculum. This chapter provides (1) a set of nine ELSI topic modules that can be incorporated into courses for both majors (from introductory to graduate level) and nonmajors and (2) examples of course pedagogy for specific classes. © 2009, Elsevier Inc.

I. INTRODUCTION

There is wide agreement, that in order to fully participate in a democratic society, a citizen must be knowledgeable and well informed (Dolan, 2008; Rutherford, 1990). An understanding of basic science is important as our society is ever more dependent on technological advances. In particular, recent advances in our understanding of the molecular mechanisms of inheritance are leading to practical applications in society, which will impact us all. Recent surveys suggest that, while the interest level is high, the general level of under-standing of genetic concepts is woefully low in the general populace (Bowling *et al.*, 2007, 2008; Dienstag, 2008). In general, a typical nonexpert not surpris-ingly acts on a need-to-know basis, becoming well educated on the mechanisms of a particular disease if they or someone close to them is affected but not becoming broadly educated in anticipation of important developments that do not directly impact their life. Recent focus on ethical, legal, and social implica-tions (ELSI) topics emerging from the Human Genome Project (HGP) has resulted in a powerful interdisciplinary mechanism for addressing this problem.

Note: the "problem" is more complex than generally recognized by scientists. It is not just that the public lacks an understanding of the "facts" of genetic mechanisms but also that scientists are poorly prepared to consider the applications or the implications of their work in society. Thus ELSI issues provide a very powerful pedagogical tool to accomplish two related goals. One is to interest nonscientists in genetics and engage them in learning the science behind the ELSI developments they are considering, and the second is to broaden the perspective of science students to consider the history and social consequences of the science they are studying. As educators we bear responsibility for ensuring our students graduate with our best understanding of the concepts, theories, and content of our disciplines. Can we honestly claim to have done that if the students are insensitive to other perspectives on this knowledge?

There has been substantial progress in the development of ELSI pedagogies, which either focus primarily on ELSI issues (the Genetics and Society type of course) or integrate these issues into "regular" genetics courses. Much of this progress is the result of the earmarking of 3–5% of the HGP budget for funding ELSI projects. One such project was the 1996–2006 Dartmouth Ethics Institute ELSI Program of workshops in which over 240 university and college faculty attended a 3–5 day workshop in preparation for implementing ELSI issues in courses at their own institution (Donovan and Green, 2008). The interest in ELSI issues as part of the undergraduate curriculum is increasing and there are many articles that provide useful resources for teaching particular topics (Haga, 2006) or offer plans to incorporate ELSI issues into a particular course (Galbraith, 2008). Our goals in this chapter are twofold: (1) to provide a set of nine ELSI topic modules that can be incorporated into courses for both majors (from introductory to graduate level) and nonmajors, and (2) to provide specific examples of course pedagogy and organization.

In Section II, two independent modules (see Sections II.A and II.B) are followed by five modules covering issues related to genetic testing and privacy (see Sections II.C–II.G). Two additional modules provide examples of more general pedagogical approaches (see Sections II.H and II.I).

Section III outlines two specific approaches to integration of ELSI issues into (A) General education courses for nonmajors and (B) Genetics courses for Bioscience majors.

II. RESOURCES

A. Eugenics

Teaching the history of a discipline can be a very effective mechanism of building an understanding of the current "state of the field." It has been standard practice in teaching genetics courses to start with Mendelian genetics and move through classical genetic analysis, to the molecular basis of inheritance (genetics and now genomics). Most genetics texts still follow this approach although the number of pages devoted to pre-1953 has been plummeting.

Over recent years another, less laudatory, chapter in the history of genetics has been drawing scholarly and public attention. Eugenics, the science of improving the human gene stock through selective breeding, was vigorously practiced during the late nineteenth and early twentieth centuries notably in the

United States and United Kingdom (Carlson, 2001). The misguided application of Mendelian genetic principles in an attempt to remove "undesirable" people from society and thus enhance society provides a powerful mechanism to alert students to the potential pitfalls in applying simple scientific principles to a poorly understood social situation (Garver and Garver, 1991). A study of the eugenics movement also illustrates how powerful and dangerous the combination of political and academic leaders with (erroneous) scientific evidence can be. A module on eugenics can be readily integrated into a standard genetics course and can be particularly useful as it fits early in the course before the background information necessary for a comprehensive understanding of molecular techniques is reached.[1] It is also a fairly easy module for an instructor uncomfortable venturing out of their disciplinary training to develop.

There are substantial resources available online. For example, the Dolan DNA Learning Center (DNALC), an operating unit of Cold Spring Harbor Laboratory, maintains an online catalog of educational resources. These include the DNALC Eugenics Archive http://www.dnai.org/e/index. html, which contains many original documents for study. A homework research assignment in which the students independently locate answers, using the documents on the CSH Web site, to questions such as "Which groups were considered to have the highest/lowest intelligence", "What was the scientific evidence used to justify this conclusion?" can form the basis for a lively discussion. During the next class students work in small groups to generate group answers to the questions and discover that some of them have slightly different answers, as there may be more than one source for a response and it depends which one they found. However, the prejudices of the time immediately become apparent and students are often startled at how brazen they seem to us.

Students tend to be very familiar with the eugenics of Nazi Germany and the Holocaust, but less so with the consequences of American eugenics, laws that affected the education, reproduction and freedom of individuals deemed to be "unfit" at that time. A short, and powerful, web exercise is to assign students to find the eugenics laws that were passed in a particular state (ideally their home state but some management is required to spread out the research), what the powers of the laws were, when they were repealed and if any official apology has been offered.

Another pedagogical approach is to focus on how different countries constructed a solution to the "problem of the unfit." The American and European approach was to try to decrease the number of inferior individuals by

---

[1] I (J. M. G.) teach eugenics during the first week of classes of the spring semester, which, in our schedule, includes Martin Luther King Day. This allows me to acknowledge the use of science in the racism of the past, and to pay respect to the work of the Civil Rights Movement in overcoming that legacy, in a science course.

isolating them, preventing them reproducing and thus preventing more being born (Carlson, 2001; D'Antonio, 2005; DNA: Pandora's Box, 2003). Ultimately the Final Solution was genocide. This approach is predicated on an us/them dichotomy in which the genetic determinism of their unfitness is never questioned and an important goal is to not allow the unfit to pollute the gene pool of the virtuous group in society.[2] A somewhat different approach, taken by the Colonial British in Australia to the "problem" of the native Aborigines, is presented in the movie "A Rabbit-Proof Fence" (Miramax, 2003). Children of mixed parentage were forcibly removed from their families, made wards of the British Governor and educated in special schools in preparation for greater integration in society. One scene, where the Governor is lecturing a group of wealthy white women sponsors on the inheritance of race, is a particularly compelling tool for use in discussion of various beliefs in the genetic basis of race. Finally, discussion of the eugenics movements of the past can set the stage for subsequent ELSI topics in a course. For example, see Section II.G.

B. Human ancestry and evolution

Students are often fascinated by a review of the history of human migration around the world and they are particularly intrigued by the use of molecular markers to trace the ancestry of human populations. The National Geographic Web site provides online access to information about the work of Spencer Wells and his colleagues to sample DNA from isolated human populations around the world. Students read *The Journey of Man, a Genetic Odyssey* (Wells, 2003) in preparation for viewing a 2-h PBS companion film at the next class meeting. Alternatively, students may view a reserve copy of the film outside of class at a campus media center. Both the book and the film are aimed at a general audience. Introductory material about DNA structure and function is appropriate for first year college students without previous genetics coursework. The use of DNA markers is explained; DNA sequence analysis is demonstrated. The book and companion film together provide an excellent opportunity to assess student familiarity with basic genetic concepts and modern DNA methods (see also Human Origins, http://www.dnai.org/d/index.html).

    The National Geographic Web site also provides information about the progress of the current Genographic Project to analyze field-collected data from indigenous and traditional peoples around the world as well as data collected from the general public. The project is anonymous, nonmedical, nonpolitical, nonprofit, and noncommercial; results of the project will be published following

---

[2]In a genetics course, one can include an interesting aside to the concept of hybrid vigor in plant breeding and how Mendelian genetics could be used to argue for such different conclusions in plants and humans in the U.S.

scientific peer review and placed in the public domain (https://www3. nationalgeographic.com/genographic; Wells, 2006). However, the collection of DNA data from Native American tribes has been problematic. Genographic origin stories told by DNA can clash with long-held beliefs vital to preserving Native American culture and may jeopardize land rights and other benefits (Harmon, 2006).

Alexander Werth has recently reviewed the implications of the Geno-graphic Project for the study of human evolution (Werth, 2008). Questions about human evolution typically arise when students are confronted with DNA evidence from comparative genomics studies. Students are baffled by reports that 90% of genes are shared between mouse and man (Waterston et al., 2002), and 99% of genes are shared between chimpanzee and man (Lander et al., 2005). Evolutionary developmental biology (Evo Devo) provides insight into the relationship between genes and evolution and is a useful point of entry to the study of evolution for students with a broad range of scientific backgrounds (Carroll, 2008; Hunter et al., 2008a). Students find the work of Sean Carroll to be accessible and engaging, regardless of their prior level of preparation in biology. In particular, Carroll's book, The Making of the Fittest (Carroll, 2006) can be useful for first and second year students enrolled in a general education course designed to fulfill a natural science in perspective requirement. Students are fascinated to learn that a small number of primitive genes led to the development of organs and appendages in all animal forms (Carroll, 2005).

Sean Carroll has two projects expected to be available in 2009. WGBH NOVA is developing a 2-h special program based on Sean Carroll's two books, Endless Forms Most Beautiful (2005) and The Making of the Fittest (2006). Carroll's new book, Remarkable Creatures: Epic Adventures in the Search for the Origins of Species, will also be published in early 2009. These should also be excellent sources for use in this module.

The potential to sample an individual's genome in order to identify that individual and determine aspects of their predisposition to health or inherited diseases gives rise to a series of five modules in issues relating to genetic testing and privacy.

C. Genetic identification: Forensics

Discussions of what is, or soon will be, possible with regard to genetic testing and individual identification can take many different approaches. Students are very familiar with the concept of individual identification through fingerprints and retinal scans. They have grown up with the various CSI series and the ubiquitous swabbing and DNA analysis that invariably contributes to a successful conclusion of each episode. Given that the technology for individual identifica-tion is well established, it is the utilization and regulation of that technology that is currently under consideration. Routine uses of genetic analysis for identifica-tion purposes include: prosecution of suspects in criminal investigations; proof of innocence in cases of wrongful conviction of people in the past (Dwyer et al., 2000; The Innocence Project—http://www.innocenceproject.org/); identification of crime or disaster victims; identification of military personnel; and proof of familial relationships (paternity and immigration cases). This type of use of genetic information is uncontroversial and has great utility; however, other applications of genetic testing and databases established for forensic purposes cause more concern even among young adults.

American society is based on a very strong concept of individual rights, carefully overseen by the American Civil Liberties Union. Privacy is established in the Constitution in various ways including the fourth Amendment offering protection against unlawful search and seizure. This sensitivity has prevented widespread universal testing of the general population around a crime scene, an approach that has been utilized to successfully apprehend criminals in many countries in Europe (McCabe and McCabe, 2008; see Chapter 11 in Mehlman et al., 2006). A role-playing exercise where all male students are required to submit samples in a fictional rape investigation quickly reveals serious tensions about submitting to such testing. Is it unfair that only men are at risk of their genetic information becoming public? Should there be genetic databases made from newborn blood samples so that everyone is already "on file"? Is it permissible for law enforcement officers to collect "abandoned" DNA samples on coffee cups and analyze them? (Once "abandoned," the coffee cup and the cells it carries are not subject to individual protection through ownership).

One useful technique to make this discussion relevant to students is to propose a college identification system (see strategies section later). Using microsatellite markers, like those used in CODIS that do not have any medically predictive value, as part of a College identification system seems like a good idea as it would allow campus security to identify the person(s) who threw a rock through the professor's office window. But what if all those beer cans on the quad after a party weekend were analyzed to catch underage revelers?

Students tend to be more comfortable than their elders with decreased privacy as members of their generation actively participate in information sharing Web sites like Facebook and YouTube. This tendency, toward informa-tion exchange as a form of social networking, is currently being exploited by genetic testing companies. The firm 23andMe, which advertises that their genetic test results give information about 80+ diseases, traits and conditions, has been running "spit parties" to promote their product suggesting that people will form social groups on the basis of the alleles they share, as in "You are invited to join the group of Slow Caffeine Metabolizers" (Salkin, 2008). A discussion based on the students' readiness to get involved with such technologies and their concerns about the privacy of such information once collected links well to the next module on genetic testing for disease related markers.

D. Genetic testing: Simple inherited diseases

This module is designed to familiarize students with online medical genetics resources and the privacy issues that arise from their use. In addition, students get the opportunity to explore relevant terms encountered in the popular media such as "personalized medicine" (Kaiser, 2008; Lee and Morton, 2008) or "genomic medicine" (Hunter *et al.*, 2008b; McBride *et al.*, 2008).

Your Genes, Your Health (YGYH) is an online resource targeted to patients and families who are looking for easy-to-understand information about a specific genetic disorder (http://www.ygyh.org). Information for each disorder is organized according to questions visitors may have about the disorder: What is it? What causes it? How is it inherited? How is it diagnosed? How is it treated? What is it like to have it? Where can I get more information? YGYH focuses on 15 disorders, which were chosen using three criteria: high incidence rate, known genetic cause, and severity of the phenotype (symptoms). The 15 disorders include Alzheimer disease, beta-thalassemia, cystic fibrosis, Down syndrome, Duchenne/Becker muscular dystrophy, Fragile X syndrome, hemochromatosis, hemophilia, Huntington disease, Marfan syndrome, neurofibromatosis, phenylketoneuria (PKU), polycystic kidney disease, sickle cell disease, and Tay–Sachs disease. In each case, the participation of genetic foundations or organizations was enlisted for information and access to patients and/or physicians for video interviews. The Web page for each disorder comprises a number of resource pages that provide in-depth information. The first "page" provides quick facts for casual browsing. Subsequent pages include detailed animations to help visitors visualize the unseen world of genes and molecules and explain the biology of the disorder. Video interviews with researchers and patients provide insiders' views on genetic disorders. Links help users find support groups and additional information.

Assignment of specific genetic diseases to each of a number of small groups of students permits students to present material to the class and share what they have learned. Alternatively, individual writing assignments can be designed to include a wider range of human genetic diseases (Burke *et al.*, 2006; Pagan, 2006).

This module can be expanded to include the study of DNA data collection projects described as Biobanks (Rothstein, 2006). These projects include the Iceland Biobank (deCODE), the UK Biobank, and The Human Genome Diversity Project. Each of those projects can be assigned to a small group of students for presentation to the class. Students may role-play as investigative reporters and opt for a news program format presentation. Alternatively, students may assume roles as scientists and participants in each of the projects, in order to demonstrate ELSI issues associated with Biobanks (Haga and Beskow, 2008).

Students are increasingly aware of press about commercial DNA services available to the public, such as 23andMe, deCODEme, and Navigenics (Pollack, 2008; Wade, 2007; Wolfberg, 2006). The Icelandic company deCODE Genetics offers a service called deCODEme, which will assess a person's genome for risk of common diseases, bodily traits like hair and eye color, and ancestral origins. A similar service is offered by 23andMe, and a third company, Navigenics, focuses on disease genes. Discussion of the regulatory action taken by California and New York against these companies (to prevent solicitation of customers in those states until an appropriate license to offer medical tests was obtained) will inform students about issues associated with the collection of personal genetic data (Wadman, 2008). Privacy issues should be explored in the context of the data collection processes associated with each of these projects (Greely, 2007; Kaiser, 2004; Lin et al., 2004; Roche and Annas, 2006).

E. Genetic testing and counseling

Until recently most genetic testing was a carefully controlled procedure conducted by medical professionals. Testing was only undertaken on individuals either affected or at risk, for an inherited disorder for which a genetic test was available. As discussed above, a new market in commercial genetic testing has developed where relatively unregulated screening of DNA isolated from saliva or cheek cell swabs, submitted directly to a company by the customer, is undertaken with no oversight by a physician or any regulatory agency. As genetic information can be powerful, affecting people beyond the individual tested, and interpreted in very different ways it is very important to consider how this information is distributed.

Discussions on the technologies and power of genetic testing (module 4) generally result in strong agreement as to the necessity of genetic privacy: that one's genome should be carefully protected under personal control as given under the American Society of Human Geneticists (ASHG) statement—"Genetic information, like all medical information, should be protected by the legal and ethical principle of confidentiality" (American Society of Human Geneticists, http://www.ashg.org/).

To introduce students to the many complications of genetic testing it is useful to get them to consider case studies that raise the ambiguities faced by genetic counselors and physicians every day. The general methodology of genetic testing can be introduced using a specific example, e.g., the breast cancer susceptibility genes *BRCA1* and *BRCA2*. Patients desiring a genetic test are interviewed by a counselor and a family history obtained. Genetic testing is most powerful when the allele associated with increased cancer risk in a particular family is known. Thus it is preferable that affected members of the family be tested first and, if a mutant *BRCA1* or *BRCA2* allele is detected, then the proband (the person requesting testing) can be tested for that allele. Results are shared with the individual at a follow-up counseling appointment.

Some scenarios for consideration:

a. A woman who has had breast cancer comes in for testing but, posttesting, decides she cannot cope with the information and does not come back for her results, which are negative. Knowing how relieved she would be with the negative result, should the counselor contact her? Months later, the woman's daughter comes in requesting testing and she intends to undergo a double mastectomy if she is found to be at increased risk. Her mother still does not know, or want to know her status, but testing the daughter will "inform" to the mother. Plus, mother's test was negative so her daughter will not gain much information from testing although that information is confidential. To whom does the counselor owe the most responsibility?

b. A woman, whose relatives have been affected with breast and ovarian cancer linked to a known mutant *BRCA2* allele, is tested. She comes in for the (negative) results and is accompanied by her sister who now also wishes to be tested but you judge to be too emotionally unstable to cope with a positive result. As the counselor do you have "right" to make that determination?

c. A woman, who is 8 weeks pregnant, is diagnosed with breast cancer and wants genetic testing. She says that if she tests positive for a mutant *BRCA1* or *BRCA2* allele she will request testing for the fetus. Your response?

A physician's or genetic counselor's responsibility to their patient for confidentiality and thus genetic privacy is sometimes in contradiction to what most people would be perceive as a greater good. For example, when a patient is diagnosed with a disease that has a well-understood pattern of inheritance (e.g., certain cancer genes) this has obvious implications for other family members, particularly any offspring of the affected individual (aggressive early screening might give significant increase in lifespan). Normally, one could expect a person to have the best interests of relatives at heart and that they would inform their relatives as appropriate so that they could all benefit from the information. However, not all families function optimally and in cases where the affected individual would not inform others of a serious risk doctor (counselor)/patient privacy may result in serious harm to another identifiable individual. Thus, through a series of test common law cases, the ASHG policy has been developed to include conditions where disclosure of private genetic information is allowed.

Disclosure is permissible. . .

- Where attempts to encourage disclosure on the part of the patient have failed
- Where serious and foreseeable harm is highly likely to occur
- Where the at-risk relative is identifiable
- Where either the disease is preventable/treatable or where early monitoring will reduce the genetic risk
- Where the harm that may result from failure to disclose outweighs the harm that may result from disclosure

After consideration of ethical concerns that can arise in the regulated arena of medical genetic testing, it is also useful to have students investigate the very different world of online genetic testing. A quick Google search for "genetic testing" will show them how easy it is to get information on certain markers; it is only a matter of some saliva or cheek cells and a few hundred dollars. For example, 23andMe claims to give your genotype at markers, including two linked to breast cancer (reading the details, these are not *BRCA1* or *BRCA2*—https://www.23andme.com/). It is interesting for students to consider how useful the results of these tests can be and what use is implied in the marketing of the tests? At present, the markers tested have very low predictive power but this situation will change rapidly as more alleles linked to "undesirable" conditions are elucidated. It has been shown that different people (personalities, ethnic backgrounds, family situations) respond very differently to risk information (Geller *et al.*, 1997). Should we regulate companies marketing genetic testing and, if so, how?

F. Genetic testing: A simulation

Simulation of genetic testing in the classroom provides an opportunity for students to reflect on the issues associated with genetic testing. This module can be accomplished in a single class meeting or expanded to include more extensive work during the semester. A wide range of genetic tests is available for simulation (http://www.myriadtests.com/inherited.htm). The module described here is designed to increase student awareness of risk for skin cancer. Online information about a test kit available from Myriad Genetics describes "Melaris", a test to determine risk for hereditary melanoma. Relevant material includes (1) Family ties and melanoma, (2) Does melanoma run in your family? (3) Inheriting a gene mutation puts you at higher risk, (4) Melaris, a test for hereditary melanoma, (5) Keeping your skin healthy, (6) Some frequently asked questions, and (7) Family history questionnaire. Each student is asked to read the brochure in class, and draft a fictional narrative based on the information distributed as part of the Melaris test kit. Students may opt to write a letter from a cancer patient to the physician explaining the decision to be tested or the decision not to be tested. Alternatively, the letter might represent the effort of a cancer patient to inform family members about the testing option and/or hypothetical testing results. This module can be supplemented with Web-based discovery of (1) genetic testing forms for informed consent and (2) state regulations about informed consent practice and procedures associated with genetic testing.

This module can be expanded to include coverage of the roles of The American Academy of Dermatology (http://www.skincarephysicians.com), the American Cancer Society (http://www.cancer.org), The Melanoma Center (http://www.melanomacenter.org/index.html), The National Cancer Institute (http://www.cancer.gov), and The National Society of Genetic Counselors (http://www.ngsc.org).

The simulation exercise can be followed by assignment of specific genetics tests to each of a number of small groups of students. Student groups can present, for example, scenarios involving a test for Huntington disease, breast cancer, or colon cancer. Alternatively, individual writing assignments can be designed to include a wide range of human diseases for which genetic testing is possible (Pagan, 2006).

Discussions of medical genetic testing regularly highlight concerns about the privacy of the information and whether other parties (employer, insurance agent, college admissions officer, potential personal or business partners) can get access to the information (Geller, 1998; Miller; 1998). Two useful videos to illustrate these points are "Do You Really Want to Know?" (CBS video, 1996) and "Bloodlines: Technology Hits Home" (Backbone Media, 2003) and either can form a basis for consideration of the following group exercise. Have students assume they are on a government subcommittee drafting regulations on genetic testing in the workplace.

1. What harm(s) are you trying to prevent occurring with these regulations?
2. What type of regulation is most appropriate? How would it be enforced?

In order for the regulations to stand up to legal challenge, you will have to be careful to define your terms. What is a genetic test? Are the newborn tests for metabolic diseases like PKU genetic? Is information about an actual genetic disease a retroactive genetic test? Could measuring height be a "genetic test" for dwarfism?

After they have addressed these questions introduce the current regulations—Health Insurance Portability and Accountability Act (HIPAA—http://www.hipaa.org/) and the Genetic Information Nondiscrimination Act (GINA—http://www.opencongress.org/bill/110-h493/show, http://www.geneticfairness.org/ginaresource.html), which became law in May, 2008, and critiques of the power and scope of these laws (Sobel, 2007). It is important to bring students to understand that we will not know the degree of protection afforded by these laws until they are tested in the courts.

G. Gene therapy and genetic enhancement

Much of the justification for the high levels of funding for the HGP was based on the potential for improvements in human health. At one level, once scientists understand the underlying mechanism of a genetic disease then it is, theoretically, much easier to design a cure. Substantial progress has been achieved on simple inherited diseases. For example, targeted therapies for cystic fibrosis are now designed with the knowledge that the underlying cause of all the symptoms is a defect in cellular chloride transport. While therapies based on an understanding of the genetic basis of a disease have not often been as successful as predicted, there is no opposition to this type of research. However, once the genetic basis for a disease is understood the potential for a genetic "cure" in the form of gene therapy or "improvement" in the form of genetic enhancement becomes apparent.

The checkered history of the promise and perils of somatic gene therapy to attempt to cure a genetic disease can be used to underscore for students how complex the physiology of multicellular organisms is and how hard simple cures are. Important topics for discussion include: why somatic gene therapy is only applicable to recessive disorders; use of vectors to target genes into cells and the related safety problems; regulation of experimentation on human subjects and informed consent. These can be focused around particular cases, for example, the attempts at gene therapy to cure adenine deaminase deficiency (ADA) in children suffering from Severe Combined Immune Deficiency Syndrome (SCIDS) which resulted in some children being later affected with clonal lymphoproliferative disorder, or the Jesse Gelsinger tragedy where 18-year-old Jesse died as a result of systemic inflammatory response syndrome during a gene therapy clinical trial (McCabe and McCabe, 2008). It is also important to emphasize to students that some of the diseases for which gene therapy was/is under development are extremely debilitating and affected children are suffering profoundly. That parents and doctors are aggressively pursuing any avenue for a cure is not surprising and their position should be carefully considered when evaluating somatic gene therapy's potential. However, somatic gene therapy is essentially under a moratorium at present while safety issues around vector choice and genome integration of inserted genes are addressed.

The inadequacies and dangers of somatic gene therapy suggest that it may never be a very useful method of treating inherited diseases, and two other mechanisms of decreasing the number of individuals affected with these diseases seem far more promising. We can either prevent affected individuals being born, through either embryo or fetus screening, or "fix" the early embryo through germ-line genetic engineering. There are multiple mechanisms currently practiced where the human embryo that is allowed to develop into a baby is selected in certain ways. Most commonly, amniocentesis and karyotyping fetal chromosomes allow the identification of fetuses with visible chromosomal abnormalities;

this allows the parents to abort that fetus. Students should consider—is prenatal genetic testing—given that many parents elect abortion if a genetic abnormality is detected—eugenics (Duster, 2003, Rifkin, 1998)? There are many interesting perspectives on this argument in the DNA: Pandora's Box video, which features parents of a child with Down Syndrome and Dr. Kay Jamieson arguing against selective abortion of children with disabilities.

Parents who are known to be carriers for a sex-linked recessive genetic disease can elect to use sperm sex sorting or *in vitro* implantation of only female embryos to ensure that they will not have an affected boy. Furthermore, the technologies developed for *in vitro* fertilization and "test-tube" babies allow for more sophisticated screening for traits. Fertilized embryos are allowed to develop to the 8-cell stage and one or two cells removed for genetic analysis. Using polymerase chain reaction (PCR) amplification of small segments of the genome, a geneticist can determine the presence of alleles associated with increased risk of a particular disorder or disease. Thus, parents can decide whether to implant a particular embryo based on the presence/absence of certain markers. When the markers screened for result in a serious genetic disease, there is no doubt that this is a very desirable strategy. It decreases the prevalence of suffering due to genetic diseases without abortion, it allows for sensible choice of the embryos for implantation—a great benefit to all concerned.

While the advantages of preimplantation screening are obvious, students are usually quick to point out two developments from these procedures that are more troubling. Firstly, if it becomes normal procedure to test for diseases, disorders, and traits that are linked to known genetic markers then which diseases, disorders, and traits are worth testing for and screening against? Students have to address the "slippery slope" arguments—after all if we are just screening embryos and choosing a few for implantation why not use the "best" as shown in GAT-TACA (Columbia, 1997). Secondly, genetically screened early embryos would be the perfect precursor for genetic enhancement where extra copies of "beneficial" genes are injected into the embryo to "improve" it. There is an excellent segment in DNA: Pandora's Box where Mario Capecchi discusses his work on artificial chromosomes in mice and then speculates about introducing extra chromosomes containing genes that can be turned on or off at will into human embryos. In that same video, they can see James Watson expressing some of his most controversial eugenic views: "People say it would be terrible if we made all girls pretty. I think it would be great." Student understanding of the issues involved can be expanded by assigning readings pro- (Caplan, 2004; Watson, 2004) and anti- (Fukuyama, 2002; McKibben, 2003; Sandel, 2004) genetic enhancement. This area is very rich in controversial topics for discussion. Can we even discuss the possibility of "genetic basis" of violence and/or risk-taking behavior without eugenic under-tones (Hamer and Copeland, 1999; Rifkin, 1998)? Can we imagine a "posthuman future" where genetic enhancement is the norm (Fukuyama, 2002; GATTACA,

1997; McKibben, 2003)? Do you still have "free will" if you are programmed to have particular abilities (McKibben, 2003)? Can there still be democracy when people are "preprogrammed" (Fukuyama, 2002)?

H. Careful communication

Communication of complex scientific ideas and experiments to a general audience is probably one of the most challenging aspects of a scientist's responsibilities. It is one for which we are poorly trained, and few of us have the rare gift of making science clear and exciting to a general audience. Articles and editorials in journals like *Science* and *Nature* regularly exhort members of the scientific community to communicate clearly with a broad audience and to invest in public education, while bemoaning the low priority given this skill. Clear concise summaries of new developments in research are particularly important when the topic under discussion has implications on social attitudes and practices. Many advances in genetics are pertinent to fundamental issues of human identity and the privilege given to scientific knowledge makes the careful dissemination of new developments all the more critical. Unfortunately, the need for conservative publication of a project may be in opposition to a very natural tendency for authors of exciting papers to make their papers accessible to a wider audience by adding a little human interest speculation to their discussion. The pitfalls of this approach are vividly illustrated in this exercise in science communication.

Distribute three articles to different groups of students.

1. Evidence from Turner's Syndrome of an imprinted X-linked locus affecting cognitive function (Skuse *et al.*, 1997)
2. A father's imprint on his daughter's thinking (McGuffin and Scourfield, 1997)
3. Genetic X-factor explains why boys will be boys (Highland, 1997)

These articles each describe the same research but are intended for different audiences. The first, a "normal" journal article reports experiments on the intelligence and social aptitude of individuals with Turner's Syndrome and compares those in whom the single X chromosome is inherited from the father with those whose X is maternally inherited. They conclude that children who inherited their single X-chromosome from their mother have a higher incidence of social difficulties despite having normal intelligence levels. This paper ends with the provocative sentence "Our data on normally developing children suggest it (imprinted gene on X-chromosome) may also exert an effect on social and cognitive abilities in the normal range." The *Nature* News and Views article in the same issue presents this paper for the general scientific audience. It generalizes and builds upon the Skuse article to present a case for the molecular basis of behavior that is gender-linked: "Now, for the first time, we have evidence about the location of a gene that plays a part in behavioral sexual dimorphism." This research was reported in the popular press (The Telegraph is a national newspaper in the UK) in the following terms "Men are born lacking a factor responsible for female intuition and social graces, says a study that reveals the first genetic basis for differences in the way that men and women behave" and continues with a series of sexist comments about undermining the trend toward sexual equality! (Highland, 1997).

One effective approach is to assign each of the articles to a group of students, have them discuss it in class, decide what they think they know from the article and what evidence there is to support their conclusions. They can then present their article to the whole group so that everyone can see how the progression of the "translation" of research for the general audience can lead to serious misrepresentation of the research. They should see that this occurs through a series of incremental changes in the meaning of the "conclusion" of the research article and tends to result from the journalists' attempts to make the article more understandable and more appealing. An important question to discuss is who, if anyone, is at fault in this case and what are the responsibilities of the scientists to prevent their research being misused in someone's social agenda.

Another pedagogical approach to achieve the same end is to charge individual students to work as investigative reporters to track down a specific primary literature article cited in a current press release found online, in a newspaper article or on broadcast television, and present the results of the investigation to the class. The scope of the problem of misrepresentation of scientific evidence in the media becomes apparent, when the class as a whole subsequently reviews the collection of articles. The discovery of misleading or sensationalized information in the media is enlightening and generates a healthy skepticism in the students.

I. Analysis of relevant fiction and media

Fiction engages student interest and is accessible to students with a broad range of backgrounds. Students can identify societal issues related to genetics by analyzing fiction and by studying the research process described by novelists. In this section, we provide examples of recent work by two novelists, Jodi Picoult and Lori Andrews. Online access to Web sites maintained by Picoult (http://www.jodipicoult.com) and Andrews (http://www.kentlaw.edu/faculty/landrews/) permits students to become acquainted with the authors and to explore the background and impact of specific novels by these authors.

Research into Vermont's eugenics project of the 1920s and 1930s was conducted by Jodi Picoult in preparation for her novel, *Second Glance* (Picoult, 2003); stem cell research and "designer babies" are issues addressed in her subsequent novel, *My Sister's Keeper* (Picoult, 2004). Teaching *My Sister's Keeper* in the undergraduate curriculum has been described in detail by Terrance McConnell (2008).

Lori Andrews is the author of three mysteries involving a fictional geneticist: *Sequence* (Andrews, 2006), *The Silent Assassin* (Andrews, 2007), and *Immunity* (Andrews, 2008). After reading and discussing one or more of these novels, students can discover a direct connection between fiction and fact, by reading nonfiction work by Andrews. Andrews is coauthor of a law school casebook (Mehlman, 2006), which documents cases relevant to genetics.

The casebook is divided into four sections. The first provides an introduction to the context in which decisions about genetics have been made. The second section covers genetic research, including issues related to federal and international regulations, intellectual property rights, and research initiatives such as the Human Genome Diversity Project and the Environmental Genome Project. The third section deals with medical applications of genetics, including prenatal testing and gene therapy. The fourth section addresses the nonmedical application of genetics, including paternity testing, and the use of genetic technologies by social institutions, including law enforcement officials, courts, insurers, employers, and schools. Individual chapters from the textbook can be assigned to small groups of students who select specific cases to dramatize for the class. For example, one student group may present a mock courtroom scene based on one or two specific patent law cases described in Chapter 5, Commercialization of Genetic Research: Property, Patents, and Conflicts of Interest. Another student group may engage in similar role-play to present a family court drama based on liability for malpractice in prenatal screening from Chapter 6, Genetic Testing and Reproduction. Students also may prefer to debate issues raised in particular cases, presenting the debate as a mock radio or television broadcast. Explicit connections between a factual case and the issues raised in one of the assigned novels can be utilized to conclude this module. In *Ferrell versus Rosenbaum* (Mehlman, 2006, p. 364), a case involving conception of a sibling to be a donor, students discover a case similar to those encountered by Jodi Picoult in her research for *My Sister's Keeper* (Picoult, 2004). Alternatively, study of individual court cases can also be organized in the context of individual writing assignments. For example, the casebook, *Andrews' Genetics: Ethics, Law and Policy*, second edition, includes detailed coverage of 36 principal cases among the 100 cases cited and discussed (Mehlman *et al.*, 2006).

Additional resources available to faculty and students for use in this, or any of the other modules, include current press reports, commercial, and educational video productions. Students can expect to find current press about genetic advances regularly in the *New York Times*, particularly in the Science Times section and Style magazine. A Pulitzer Prize-winning series, The DNA Age, by Amy Harmon is also available online from the *New York Times* (http://topics.

nytimes.com/top/news/national/series/dnaage/index.html). The frequent availability of late-breaking news from these sources improves student awareness of
the rapid pace of genetic advances. It is useful to ask the students to share press
releases from other online, broadcast or print news sources with the class for
regular discussion of ELSI issues raised by these advances. Catalogs of educational video productions are available from the Howard Hughes Medical
Institute (http://www.holidaylectures.org). For example, the 2002 lecture set
entitled Scanning Life's Matrix: Genes, Proteins, and Small Molecules includes
two lectures by Eric Lander covering genetic advances, comparative genomics,
and genetic variation. Use of this material in a nonmajors course effectively
provides expert guest lectures to supplement course material. Useful videos
can also be purchased from Films for the Sciences and Humanities (http://
www.films.com). Additional film resources are described in Sections II.A, II.B,
II.F, and II.G.

III. STRATEGIES

A. Integration of ELSI issues into nonmajors courses

Many undergraduate institutions offer nonmajors topics courses, some as part of
distribution or seminar programs but all with the goal of providing a broad
education to their students. These courses offer unique opportunities to teach
about human genetics and its consequences to a broad spectrum of students;
however, they also provide substantial challenges as an instructor may be required to find common ground among students with no college-level science
preparation and those currently enrolled in sophomore-level college Biology and
Chemistry coursework. Each of the authors teaches one of these ELSI courses at
their home institutions in typical one semester 3 h/week format.

The goal of Genetic Testing, the ELSI course taught by K.L.T., is to
examine genetic testing from a variety of scientific and societal perspectives. The
context for the course is the natural world, in which all living creatures share the
universal language encoded in DNA. The focus of the course is the DNA
fingerprint, a genetic test that can be applied to DNA from any organism.
The class is divided into groups of no more than five students, when group
work is appropriate. The instructor divides students into groups to guarantee
that each group includes at least one student with a strong college-level science
background. For example, a group ideally might include a premedical student, a
prelaw student, a student majoring in business, a student majoring in the
arts, and a student majoring in the social sciences. Group work includes planning sessions during and outside class time to prepare for group oral presentations. Typically students elect to divide the presentation of material by

roles to be played in a mock trial, radio or television broadcast, or by debating teams. Groups are challenged to present as many different perspectives on a particular issue as possible. Within a group, students are also challenged to advocate for a position that does not necessarily reflect their own personal point of view.

The semester begins with the introduction of DNA fingerprint technology in the context of current studies in human ancestry and evolution (see Section II.B). Next, group work focused on DNA biobanks culminates in class presentations. In parallel, written responses to weekly reading assignments about evolution are collected from students individually. Preparation for a second round of group class presentations covering the use of DNA fingerprint technology in medical diagnostics (see Section II.D) follows simulation of a medical genetic test in the classroom (see Section II.F). The second half of the semester is devoted to a deeper investigation of ELSI issues based on relevant fiction, media, and court cases (see Section II.I). The final project for the course is a significant paper challenging each student to demonstrate their ability to communicate results and implications of a unique genetic study based on information from a wide range of sources, including a primary literature reference and press releases about that work available from the popular press.

J.M.G. teaches a Genetics and Society course, which focuses on genetic determinism and issues of human identity. This course evolved from an interdisciplinary seminar, which was co-taught with a philosopher, and interdisciplinarity is a common denominator for the following examples of courses that also could be adapted to courses for nonmajors. Beecher-Monas (2008) has described a law school seminar course, Genetics and Law, on future dangerousness predictions in the courts. Students explore genetic predictions for violent behavior, material that could be adapted for an interdisciplinary general education course based in psychology and sociology. Segady (2008) has described an interdisciplinary course based in bioethics and sociology, Ethical Futures: Implications of the Human Genome Project, which could be also adapted for a general education curriculum. Stober and Yarri (2008) developed a team-taught course, God, Science, and Designer Genes, integrating the disciplines of biology and Christian theology, which was designed to fulfill either general education biology or ethics requirements.

B. Integration of ELSI issues into genetics courses for bioscience majors

In a recent survey of professors teaching undergraduate genetics courses, most believed that it is very important to include the teaching of ELSI issues in their courses (Booth and Garrett, 2004). However, these same instructors almost unanimously indicated they had insufficient time to adequately cover ELSI issues in their courses. Thus, it is very important to devise mechanisms of integrating ELSI discussions into regular coursework with minimal loss of the "core" material. This works best when the ELSI topic is closely aligned with the syllabus. As discussed earlier, the module on eugenics fits well into the first weeks of a genetics course when Mendelian and classical human genetics are discussed. One or more of the genetic testing modules can be incorporated as the class learns about the techniques of molecular biology and explores the potential of genomics. One way of avoiding the loss of class time is to utilize lab time. For example, if the students are running a genetic testing laboratory exercise (e.g., microsatellite analysis of cheek cell DNA) then one lab period could be spent isolating DNA, setting up the PCR and watching one of the videos about ELSI issues arising from genetic testing. During the next laboratory, they can discuss an ELSI topic while the gel of their PCR samples is running. It is particularly beneficial for them to discuss the applications and implications of a technology at the same time as they directly experiencing the challenges of getting good results themselves.

Another approach involves the assignment of "new" College ID cards simulating a time (in the not-too-distant future) when individual genomes can be readily analyzed and/or sequenced so that particulars of an individual's genome are present on the card. This "information" can take a variety of forms and be used in several ways. Students assigned high risk for a disease or trait can consider how this information is useful in terms of prevention/family planning, etc. or whether the knowledge negatively impacts their self-image. Or students who "discover" that they have been genetically enhanced with extra DNA realize the potential of preimplantation genetic enhancement. The genetic identity card approach gives the advantage of focusing the students on a "real" situation. When in a hypothetical discussion most students tend to reject enhancement as too freakish. However, once they realize that they "know" that they have an increased risk of colon cancer or early memory loss and dementia, they start to want to do something about it. And, when there is the possibility of a genetic improvement (e.g., an artificial chromosome with extra "good" copies of HNPCC1 to decrease the likelihood of cancer), then they start to waiver. If the class is deliberately engineered where some of them have genetic enhancement chromosomes and others do not, the discussion usually becomes pro-enhancement because, in our competitive society, everyone wants full access to what is available, exactly the genetic one-upmanship discussed by McKibben in his book criticizing these technologies, *Enough* (McKibben, 2003).

The genetic identity card can be used in many ways. It can form the foundation for an entire nonmajors general education course (Section II.A) in genetic testing and determinism. Or a card can be used to generate outside class discussion among bioscience majors and their nonscience peers. For example, students who have cards could lead a discussion at a student club meeting or after the showing of a futuristic film, for example, GATTACA. Thus majors are encouraged to both engage the ELSI issues arising from their studies and enhance their communication skills with the "public," both highly desirable outcomes for our future scientists.

IV. CONCLUDING REMARKS

The application of science and technology in society often leads to important and controversial decisions that must be resolved by both scientists and the public. Among the decisions that have been/are being made are those that have the potential to fundamentally alter life as we know it. For example, nuclear technology with its potential to provide massive amounts of cheap energy, which is accompanied by potential for intentional (weapons) or unintentional (accidents) destruction, has been a source of controversy for decades. Mishandled or misregulated, nuclear technologies clearly have the potential for catastrophe. Less dramatic, but no less fundamentally serious, is the rate of degradation of the natural environment and the vexing problems of balancing human development with preservation of the natural world. The emerging technologies in genetics and neuroscience also have the potential to alter humanity in fundamental ways we are only just beginning to address (Fukuyama, 2002; McKibben, 2003). It is very important that we educate students to become scientifically literate citizens, to give them the knowledge and understanding they need to engage effectively in discussions, debates, and decision-making involving science in their personal and professional lives, and to help them appreciate science and the roles it plays in modern society. The resources and strategies presented in this chapter for teaching ELSI issues that arise in modern genetics are designed to aid in accomplishing this goal throughout the undergraduate curriculum.

Acknowledgments

Both authors of this chapter attended a Dartmouth Ethics Institute ELSI workshop and gratefully acknowledge the contribution that material made to their development of the ELSI exercises outlined here. Special thanks from J.M.G. to Jennifer Axilbund, Cancer Risk Assessment Program, Johns Hopkins Hospital for her presentation to the Dartmouth Ethics Institute ELSI Workshop.

References

Andrews, L. (2006). Sequence. St Martin's Press, New York, NY.
Andrews, L. (2007). The Silent Assassin. St Martin's Press, New York, NY.
Andrews, L. (2008). Immunity. St Martin's Press, New York, NY.

Beecher-Monas, E. (2008). Teaching genomics and law by exploring genetic predications of future dangerousness: Is there a blueprint for violence? In "The Human Genome Project in College Curriculum" (A. Donovan and R. M. Green, eds.), pp. 132–144. University Press of New England, Hanover, NH.

Booth, J. M., and Garrett, J. M. (2004). Instructors' practices and attitudes toward teaching ethics in the genetics classroom. *Genetics* **168,** 1111–1117.

Bowling, B. V., Huether, C. A., and Wagner, J. A. (2007). Characterization of human genetics courses for nonbiology majors in U.S. colleges and universities. *CBE Life Sci. Educ.* **6,** 224–232.

Bowling, B. V., Acra, E. E., Wang, L., Myers, M. F., Dean, G. E., Markle, G. C., Moskalik, C. L., and Huether, C. A. (2008). Development and evaluation of a genetics literacy assessment instrument for undergraduates. *Genetics* **178,** 15–22.

Burke, W., Khoury, M., Stewart, A., and Zimmern, R. (2006). The path from genome-based research to population health: Development of an international public health genomics network. *Genet. Med.* **8,** 451–458.

Caplan, A. L. (2004). What's morally wrong with eugenics? In Health, Disease, and Illness: Concepts in Medicine. Georgetown University, Washington, DC.

Carlson, E. A. (2001). The Unfit: A History of a Bad Idea. Cold Spring Harbor Laboratory Press, New York.

Carroll, S. B. (2005). Endless Forms Most Beautiful. W.W. Norton and Co., New York.

Carroll, S. B. (2006). The Making of the Fittest. W.W. Norton and Co., New York.

Carroll, S. B. (2008). Evo-devo and an expanding evolutionary synthesis: A genetic theory of morphological evolution. *Cell* **134,** 25–36.

D'Antonio, M. D. (2005). The State Boys Rebellion. Simon & Schuster, New York.

Dienstag, J. L. (2008). Relevance and rigor in premedical education. *N. Engl. J. Med.* **359,** 221–224.

Dolan, E. (2008). Recent research in science teaching and learning. *CBE Life Sci. Educ.* **7,** 171–172.

Donovan, A., and Green, R. M. (2008). The Human Genome Project in College Curriculum. Dartmouth College Press, University Press of New England, Lebanon, NH.

Duster, T. (2003). Backdoor to Eugenics. 2nd edn., Routledge, NY.

Dwyer, J., Neufield, P., and Scheck, B. (2000). Junk science. In Actual Innocence: Five Days to Conviction and Other Dispatches from the Wrongly Convicted. Doubleday, New York pp. 158–171.

Fukuyama, F. (2002). Our Posthuman Future. Farrar, Strauss & Giroux, New York.

Galbraith, A. (2008). Integrating ELSI concepts into upper division science courses. In "The human genome project in college curriculum" (A. Donovan and R. M. Green, eds.). Dartmouth College Press, University Press of New England, Lebanon, NH.

Garver, K., and Garver, B. (1991). Eugenics: Past, present & future. *Am. J. Hum. Genet.* **49,** 1109–1118.

Geller, L. (1998). Current developments in genetic discrimination. In "The Double-Edged Helix: Social Implications of Genetics in a Diverse Society" (J. S. Alper, ed.). Johns Hopkins University Press, Baltimore, MD.

Geller, G., Strauss, M., Bernhardt, B. A., and Holtzman, N. A. (1997). Decoding informed consent: Insights from women regarding breast cancer susceptibility screening? *Hastings Cent. Rep.* **27,** 28–33.

Greely, H. T. (2007). The uneasy ethical and legal underpinnings of large-scale genomic biobanks. *Ann. Rev. Genom. Human Genet.* **8,** 343–364.

Haga, S. B. (2006). Teaching resources for genetics. *Nat. Rev. Genet.* **7,** 223–229.

Haga, S. B., and Beskow, L. M. (2008). Ethical, legal, and social implication of biobanks for genetics research. *Adv. Genet.* **60,** 505–544.

Hamer, D. H., and Copeland, P. (1999). Living with Our Genes. Anchor Books, New York.

Harmon, A. (2006). DNA gatherers hit snag: The tribes don't trust them. *New York Times*, December 10, 2006.

Highland (1997). Genetic X-factor explains why boys will be boys. *The Weekly Telegraph*, June 12, 1997.

Hunter, D. J., Altshuler, D., and Rader, D. J. (2008a). From Darwins's finches to canaries in the coal mine—Mining the genome for new biology. *N. Eng. J. Med.* **358,** 2760–2763.

Hunter, D. J., Khoury, M. J., and Drazen, J. M. (2008b). Letting the genome out of the bottle—Will we get our wish? *N. Eng. J. Med.* **358,** 105–107.

Kaiser, J. (2004). Privacy rule creates bottleneck for U.S. biomedical researchers. *Science* **305,** 168–169.

Kaiser, J. (2008). Departing U.S. Genome Institute director takes stock of personalized medicine, Francis Collins interview. *Science* **320,** 1272.

Lander, E., Mikkelsen, T. S., Hillier, L. W., Eichler, E. E., Zody, M. C., Jaffe, D. B., Yang, S.-P., Enard, W., Hellmann, I., Lindblad-Toh, K., Altheide, T. K., Archidiacono, N., *et al.* (2005). The Chimpanzee Sequencing and Analysis Consortium, Initial sequence of the chimpanzee genome and comparison with the human genome. *Nature* **437,** 69–87.

Lee, C., and Morton, C. C. (2008). Structural genomic variation and personalized medicine. *N. Eng. J. Med.* **358,** 740–741.

Lin, Z., Owen, A., and Altman, R. B. (2004). Genomic research and human subject privacy. *Science* **305,** 183.

McBride, C. M., Alford, S. H., Reid, R. J., Larson, E. B., Baxevanis, A. D., and Brody, L. C. (2008). Putting science over supposition in the arena of personalized genomics. *Nat. Genet.* **40,** 1293–1296.

McCabe, L. L., and McCabe, E. R. B. (2008). DNA: Promise and Peril. University of California Press, Berkeley, CA.

McConnell, T. (2008). Best interests and my sister's keeper. *In* "The Human Genome Project in College Curriculum" (A. Donovan and R. M. Green, eds.), pp. 34–45. University Press of New England, Hanover, NH.

McGuffin, P., and Scourfield, J. (1997). A father's imprint on his daughter's thinking. *Nature* **387,** 652–653.

McKibben, B. (2003). Enough: Staying Human in an Engineered Age. Times Books, New York.

Mehlman, M. J. (2006). Forensics: Genetic testing for identification. *In* "Andrews' Genetics: Ethics, Law and Policy" (M. J. Mehlman, L. B. Andrews, and M. A. Rothstein, eds.), 2nd edn., pp. 662–712. West Publishing American Casebook Series, Salt Lake City, UT.

Miller, P. S. (1998). Genetic discrimination in the workplace. *J. Law Med. Ethics* **26,** 189–197.

Pagan, R. A. (2006). GeneTests: An online genetic information resource for health care providers. *J. Med. Libr. Assoc.* **94,** 343–348.

Picoult, J. (2003). Second Glance. Washington Square Press, New York.

Picoult, J. (2004). My Sister's Keeper. Washington Square Press, New York.

Pollack, A. (2008). DNA profile provider is cutting its prices. *New York Times*. September 9, 2008.

Rifkin, J. (1998). The Biotech Century. Tarcher/Putnam, New York.

Roche, P. A., and Annas, G. J. (2006). DNA testing, banking and genetic privacy. *N. Eng. J. Med.* **355,** 545–546.

Rothstein, M. A. (2006). Expanding the ethical analysis of biobanks. *In* "Andrews' Genetics: Ethics, Law and Policy" (M. J. Mehlman, L. B. Andrews, and M. A. Rothstein, eds.), 2nd edn., pp. 133–138. West Publishing American Casebook Series, New York.

Rutherford, F. J. (1990). Science for all Americans, Project 2061. American Assoc. for the Advancement of Science Oxford University Press, New York.

Salkin, A. (2008). When in doubt, spit it out. *New York Times*, September 14, 2008.

Sandel, M. J. (2004). The case against perfection. *Atl. Mon.* **292,** 50–62.

Segady, T. (2008). Teaching Ethical Futures; Implications of the Human Genome Project. *In* "The Human Genome Project in College Curriculum" (A. Donovan and R. M. Green, eds.), pp. 88–99. University Press of New England, Hanover, NH.

Skuse, D. H., James, R. S., Bishop, D. V. M., Coppin, B., Dalton, P., Aamodt-Leeper, G., Bacarese-Hamilton, M., Creswell, C., McGurk, R., and Jacobs, P. A. (1997). Evidence from Turner's Syndrome of an imprinted X-linked locus affecting cognitive function. *Nature* **387**, 705.

Sobel, R. (2007). The HIPAA paradox: The privacy rule that's not. *Hastings Cent. Rep.* **37**, 40–50.

Stober, S., and Yarri, D. (2008). God, science and designer genes: An interdisciplinary pedagogy. *In* "The Human Genome Project in College Curriculum" (A. Donovan and R. M. Green, eds.), pp. 100–121. University Press of New England, Hanover, NH.

Wade, N. (2007). Experts advise a grain of salt with mail-order genomes, at $1000 a pop. *New York Times*. November 17.

Wadman, M. (2008). Gene-testing firms face legal battle. *Nature* **453**, 1148–1149.

Waterston, R. H., Lindblad-Toh, K., Birney, E., Rogers, J., Abril, J. F., Agarwal, P., Agarwala, R., Ainscough, R., Alexandersson, M., An, P., Antonarakis, S. E., Attwood, J., *et al.* (2002). Mouse Genome Sequencing Consortium. Initial sequencing and comparative analysis of the mouse genome. *Nature* **420**, 520–562.

Watson, J. D. (2004). DNA: The Secret of Life. Knopf, New York.

Wells, S. (2003). Journey of Man. Princeton University Press, Princeton, NJ.

Wells, S. (2006). Deep Ancestry: Inside the Genographic Project. National Geographic Society, Washington, DC.

Werth, A. (2008). The Human Genome Project: Implications for the study of human evolution. *In* "The Human Genome Project in College Curriculum" (A. Donovan and R. M. Green, eds.), pp. 73–87. University Press of New England, Hanover, NH.

Wolfberg, A. J. (2006). Genes on the web—Direct-to-consumer marketing of genetic testing. *N. Eng. J. Med.* **355**, 543–545.

Media

Backbone Media (2003). *Bloodlines: Technology Hits Home.* PBS documentary. Producer & director Noel Schwerin. (Burlington Railroad worker's discrimination case video).

CBS video (1996). *Do You Really Want to Know?* 60 Minutes CBS Video. 4/21/96 (test for Huntington's disease video).

Columbia Pictures (1997). *GATTACA.* Producers, Danny DeVito, Michael Shamberg, Stacey Sher; director, Andrew Nichol.

Films for the Humanities & Sciences (2003). *DNA: Pandora's Box.* Series producer, David Dugan; editor, Joe Bini.

Howard Hughes Medical Institute's (2002). *Scanning Life's Matrix: Genes, Proteins and Small Molecules.* Holiday Lectures on Science.

Miramax Home Entertainment (2003). *Rabbit-Proof Fence.* Burbank, CA: Buena Vista Home Entertainment. Producers, Phillip Noyce, Christine Olsen, John Winter; director, Phillip Noyce.

PBS Documentary (2003). *The Journey of Man, a Genetic Odyssey.*

Web sites

23andMe. http://www.23andme.com/.

The American Academy of Dermatology. http://www.skincarephysicians.com.

The American Cancer Society. http://www.cancer.org.

American Society for Human Genetics. http://www.ashg.org.

Coalition for Genetic Fairness. http://www.geneticfairness.org/ginaresource.html.

The DNA Age. http://topics.nytimes.com/top/news/national/series/dnaage/index.html.

Dolan DNA Learning Center. http://www.dnalc.org.

Films for the Humanities & Sciences. http://www.films.com.

Genetic Fairness. http://www.geneticfairness.org/ginaresource.html.

Genetic Information Nondiscrimination Act (GINA). http://www.opencongress.org/bill/110-h493/show.

Genographic Project. https://www3.nationalgeographic.com/genographic.

Health Insurance Portability and Accountability Act (HIPAA). http://www.hipaa.org/, http://www.geneticfairness.org/ginaresource.html.

Howard Hughes Medical Institute. Holiday Lectures on Science. http://www.holidaylectures.org.

Human Origins. http://www.dnai.org/d/index.html.

Innocence Project. http://www.innocenceproject.org/.

Jodi Picoult. http://www.jodipicoult.com.

Lori Andrews. http://www.kentlaw.edu/faculty/landrews/.

The Melanoma Center. http://www.melanomacenter.org/index.html.

Myriad Genetic Tests. http://www.myriadtests.com/inherited.htm.

The National Cancer Institute. http://www.cancer.gov.

The National Society of Genetic Counselors. http://www.nsgc.org.

Vermont Eugenics, a Documentary History. http://www.uvm.edu/~eugenics/vtsurvey.html.

Your Genes, Your Health. http://www.ygyh.org.

3

Synthetic Genetic Interactions: Allele Dependence, Uses, and Conservation

Joseph V. Gray and Sue A. Krause
Molecular Genetics and Integrative & Systems Biology, Faculty of Biomedical and Life Sciences, University of Glasgow, Glasgow, United Kingdom

Advances in Genetics, Vol. 66
Copyright 2009, Elsevier Inc. All rights reserved.

0065-2660/09 $35.00
DOI: 10.1016/S0065-2660(09)66003-X

ABSTRACT

Genetic interactions occur between a pair of genes when the phenotype of the double mutant leads to an unexpected phenotype, one that is not predicted from the phenotypes of the single mutants alone. Here, we focus on genetic enhancements, otherwise known as synthetic genetic interactions, where the double mutant phenotype is more severe than expected. Such interactions are rife in natural populations and underlie complex traits, variable penetrance, variable expressivity, and genetic predisposition. Such interactions can also contribute valuable information for functional genomics analysis.

Pairwise synthetic genetic interactions are now being systematically uncovered for some simple model genomes. These data are affording us an unparalleled opportunity to examine, understand and exploit genetic enhancements. Here we focus on some key lessons, insights, and confusions arising from these large-scale datasets. We consider if genome-wide datasets support traditional assumptions about the functional relationships between gene products that underlie genetic enhancements. We argue that the genetic enhancement network of an organism is not uniform in nature and is highly dependent on the nature of the interacting alleles. We consider how such genetic networks can be exploited to inform gene product function. Finally, we consider the extent to which genetic enhancement networks are conserved between species.   © 2009, Elsevier Inc.

I. GENETIC ENHANCEMENT

A. Background

Genetic interactions occur between alleles at different loci and they are, by their nature, surprising. The double mutant may display a phenotype much weaker than would be expected (genetic suppression or epistasis) or much stronger than would be expected (genetic enhancement) (e.g., see Boone *et al.*, 2007; Guarente, 1993; Hartman *et al.*, 2001; Ooi *et al.*, 2006). Here we will focus on genetic enhancements, also known as synthetic genetic interactions. These interactions are being discovered systematically and in large numbers in genome-wide screens in some simple model organisms and thus dominant the known genetic interaction network. Second, these interactions are biologically very important and underpin many complex, multigenic traits (Hartman *et al.*, 2001).

Multigenic traits have a long history in the study of genetics, being considered by Johannsen and later by East at the turn of the twentieth century. However, Altenburg and Muller's work on truncate flies provided the first concrete examples of pairwise genetic enhancements, in these cases between a mutation on chromosome 2 and mutations on chromosome 1 or chromosome 3 (Altenburg and Muller, 1920).

All genetic enhancements are examples of complex multigenic traits. However, not all complex traits are examples of genetic enhancement. For example, many loci contribute to quantitative traits but with no associated enhancement of phenotype. In such cases, phenotype is determined by the cumulative effect of many mutations at many loci, the mere sum of the parts.

B. Definition

A genetic enhancement interaction is specified by the deviation of the double mutant phenotype from that expected in the absence of any genetic interaction. Predicting the latter is central to determining if a genetic enhancement is occurring or not.

Consider the case of two separate mutations a and b at loci A and B respectively. Further, the fitness (f) of each mutant can be quantified. If wild-type A/A B/B elephants have a fitness of 1 ($f_{wt} = 1.0$), let us consider the case where homozygote mutant a/a B/B has a fitness of 0.9 ($f_a = 0.9$) and homozygote mutant A/A b/b has a fitness of 0.8 ($f_b = 0.8$). The fitness defect caused by each mutation would be $1 - f$ or 0.1 for mutation a and 0.2 for mutation b. What double mutant fitness (strain $a/a\,b/b$) do we expect if genes A and B do *not* genetically interact? Here are a few reasonable possibilities with numerical results calculated for our test case:

$$f_{a,b} = \text{average of } f_a \text{ and } f_b \text{ is } (f_a + f_b)/2 = 0.85, \qquad (3.1)$$

$$f_{a,b} = \text{the worst of } f_a \text{ and } f_b \text{ is } f_b = 0.8, \qquad (3.2)$$

$$f_{a,b} = \text{consequence of additive defects is } 1 - (1 - f_a) - (1 - f_b) = 0.70, \quad (3.3)$$

$$f_{a,b} = \text{the product of } f_a \text{ and } f_b \text{ is } f_a \times f_b = 0.72. \qquad (3.4)$$

The first case requires mutation a to somehow make the double mutant healthier than mutant b alone. This would be an example of genetic suppression, not the lack of a genetic interaction (e.g., see St Onge et al., 2007). The second case is more reasonable, but here mutation b is preventing mutation a from having any effect on phenotype. This would be an example of epistasis, not the absence of a genetic interaction (e.g., see St Onge et al., 2007). The third and fourth cases are most reasonable and indeed both have been used in quantitative analysis. Intuitively, the fourth case, where $f_{a,b} = f_a \times f_b$ is the most satisfying and is gaining broad acceptance in the field: each mutation is affecting fitness to the same fractional extent whether in a wild-type background or in the presence of the other mutation (e.g., see St Onge *et al.*, 2007). For example, mutation *b* reduces fitness of wild-type cells to 0.8, but it also reduces the fitness of mutant *a* ($f_a = 0.9$) to 0.8 of its level (0.9 × 0.8).

In the absence of any genetic interaction, we therefore expect the double mutant to behave as follows:

$$f_{a,b} = f_a \times f_b.$$

However, it should be noted that the alternative "sum of the defects" model, where $f_{a,b} = 1 - (1 - f_a) - (1 - f_b)$ is in practice a very close approximation, deviating from $f_a \times f_b$ by a fitness of 0.02 in our test case.

We now have our benchmark to quantitatively measure a genetic interaction:

$$\text{Genetic supperssion} : f_{a,b} > f_a \times f_b,$$

$$\text{Genetic enhancement} : f_{a,b} < f_a \times f_b.$$

The stronger the observed phenotype deviates from the expected phenotype, the stronger the genetic interaction. However, current large-scale analyses tend to treat the existence of an interaction as the key observation, not its strength.

Genetic enhancements can involve a whole range of possible interacting alleles. The alleles could be gain of function (gof) or loss of function (lof). They could be present as the sole alleles (haploid mutant or homozygous diploid mutant) or as heterozygotes. We might expect the propensity to take part in genetic enhancements, and possibly the very nature of the interaction is dependent on the allele type. We return to this point later.

We also expect different phenotypes to affect the propensity to take part in genetic enhancements. Phenotypes can vary from the very specific and thus likely to be specified by a small number of processes, for example, eye size of *Drosophila*, to the more general that are influenced by many different processes, for example, organism growth rate or lifespan. We expect that the more general the phenotype, the greater the number of possible genetic enhancements across the genome; the more specific the phenotype, the fewer.

Although examples of genetic enhancements involving many different allele and phenotype combinations have been reported, most systematic effort has focused on identifying and studying pairwise combinations of lof mutations that cause apparent lethality or slow growth (e.g., Pan *et al.*, 2004; Tong *et al.*, 2001, 2004). We will focus on these interactions hereafter.

Synthetic lethal interactions require a large difference in observed fitness between the double and single mutants. In the case of haploid yeast strains, the single mutants form a colony; the double mutants do not. In this case $f_{a,b} < f_a$ or f_b or $f_a \times f_b$, indicating robust genetic enhancement between the interacting loci. However, double mutants often grow, but more slowly than expected. Synthetic lethal and synthetic slow growing interactions can be considered as part of a single continuum interactions differing only in relative strength. Indeed, many double mutant phenotypes classified as lethal may simply be very slow growing. These interactions are often considered together as synthetic sick or lethal (SSL) interactions or simply under the umbrella term of "synthetic lethal" interactions. We adhere to this latter convention here unless stated otherwise.

II. EXPECTED UNDERLYING FUNCTIONAL RELATIONSHIPS

In considering the functional relationships between gene products that might underlie genetic enhancements, we will focus on pairwise genetic enhancements occurring between lof mutations, the dominant cases in known genetic networks. Because it has long been postulated that the functional basis for genetic enhancements is dependent on the nature of the interacting alleles (e.g., see Guarente, 1993), we will consider different classes of lof allele separately.

In all cases, we assume that the gene products act in discrete linear functional pathways, where the activity of each local pathway is completely dependent on each of its components (e.g., of a signal transduction pathway). Furthermore, each gene product acts only within its respective pathway, that is, it does not have any additional, independent functions.

A. Interactions between null alleles

Null mutations render a gene completely nonfunctional. In the case of yeast deletion mutants, the entire open reading frame is missing, with no possibility of any trace residual function remaining (Winzeler et al., 1999). A null–null genetic enhancement could not occur between two genes in the same pathway since each mutation alone compromises pathway activity (Ooi et al., 2006). In this case $f_{a,b} = f_a = f_b$, an example of colinear epistasis, not genetic enhancement. Hence, null–null genetic enhancements can only occur between genes encoding components in *different* pathways (Ooi et al., 2006) (Fig. 3.1A). We therefore expect networks of null–null genetic enhancements to be underpinned by parallel functional relationships.

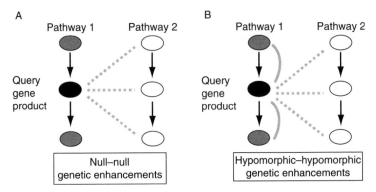

Figure 3.1. The functional relationships underlying genetic enhancements. Gene protein products (circles) act in two independent pathways that execute a common or buffering function. Each pathway is linear with the activity of each protein being fully dependent on the activity of the protein above it (indicated by black arrows). The activity of each pathway is dependent on the function of each of its components. We consider the genetic enhancement interactions (gray lines) elicited by "query" mutations affecting the gene corresponding to the black "query" protein. (A) All the interacting alleles are nulls. Genetic enhancements can only occur between pairs of mutations that affect the two parallel pathways (dashed gray lines). (B) All the interacting alleles are hypomorphic. Each mutation only partly disables the activity of the relevant pathway. Genetic enhancements can now occur between genes encoding components of the two pathways (dashed lines) or can also occur between genes encoding components of the same pathway (filled gray lines).

A genetic enhancement requires a dramatic, synergistic effect on phenotype when two mutations are combined. Such synergy is a natural consequence of parallel, redundant, or buffering relationships: each gene product buffers the organism to the full phenotypic consequences of loss of the other. Only loss of both gene products causes a catastrophic loss of biological function with a consequent dramatic effect on overall phenotype. However, it should be noted that not all genetic enhancements might be underpinned by a synergistic effect at the level of direct biological activity, such as the amount of ribosome activity in a cell. Apparent synergism can also occur at the level of phenotype if, for example, some minimal threshold of the biological activity affected by the mutations is required for the normal phenotype and only the double mutant fails to achieve this threshold.

Overall, null–null genetic enhancements should be underpinned by parallel functional relationships. In reality, there will be some exceptions, where genetic interactions occur between genes encoding components of the same pathway/process. These "within pathway" interactions arise because many *in vivo* pathways and processes are not linear, but branched. However, the genetic enhancements do not occur because the gene products act in a pathway. Rather they occur because the gene products again play independent, buffering roles. A good example is the synthetic lethal interaction between the yeast MAP kinase gene *SLT2/MPK1* and the cell cycle-regulated transcription factor gene *SWI4* (Gray et al., 1997). This appears to be a "within pathway" interaction because Slt2/Mpk1 directly activates Swi4 (Madden et al., 1997) as part of the cell wall integrity pathway that prevents cell lysis (Levin, 2005). However, the two gene products also play unique roles that contribute to the enhancement of phenotype noted for the double mutant: Slt2/Mpk1 activates other targets independently of Swi4 to promote cell wall integrity (e.g., Rlm1: Watanabe et al., 1997), while Swi4 can be activated independently of Slt2 by cyclin-dependent protein kinase (Cdc28) to express genes involved in cell wall synthesis (Koch et al., 1996).

B. Interactions between hypomorphic alleles

A hypomorphic allele lowers but does not abolish the activity of the encoded gene product. What relationships should underpin genetic enhancement interactions between hypomorphic alleles? In principle, parallel interactions reminiscent of null–null interactions could also occur between mutations affecting independent buffering functions (Fig. 3.1B). However, hypomorphic alleles present an additional possibility not available to null alleles: that pairs of hypomorphic alleles affecting the same linear pathway can also interact (Guarente, 1993; Hartman et al., 2001) (Fig. 3.1B). Each mutation alone reduces pathway activity, but the two mutations together cause further compromise in activity that results in an enhanced phenotype. Importantly, such "within pathway" genetic enhancements are now due to the effects on the shared roles of the two interacting genes, not to any distinct roles. Indeed, hypomorphic mutations have long been thought to efficiently sensitize an organism to second mutations affecting the same pathway or biological process, with many supporting *ad hoc* observations in a wide range of experimental systems, including *S. cerevisiae*, *C. elegans*, and *D. melanogaster* (e.g., see Guarente 1993; Hartman et al., 2001; St Johnston 2002).

Why would the combination of two hypomorphic mutations affecting the same pathway result in genetic enhancement? There must be synergy between the mutations at the level of pathway activity or phenotype or both. Synergy at the level of the pathway is not automatically guaranteed for a simple linear pathway. However, synergy between the mutations can occur at the level of the pathway if for example the pathway is under negative feedback control. Alternatively, synergy can occur at the level of phenotype if there is a threshold of pathway activity below which phenotype is dramatically affected.

Our consideration of genetic interactions involving hypomorphic mutations should not be allele specific as such. Any allele or genotype that reduces gene product activity is relevant. Alleles can be inherently hypomorphic, for example a temperature-sensitive allele at semipermissive temperature. Even null alleles can cause a hypomorphic effect on gene activity, for example, when present in the heterozygous state.

C. Interactions between hypomorphic and null alleles

Hypomorphic–null interactions should share characteristics of both hypomorphic–hypomorphic and null–null interactions. The vast majority of detected hypomorphic–null interactions occur between essential and nonessential genes, respectively. From the standpoint of the essential gene partner, we might expect enrichment for second site mutations that affect the same function or pathway—"within pathway" interactions (Fig. 3.2). Indeed, essential proteins often act as hubs with networks of nonessential proteins feeding into and out of them (Jeong et al., 2001). In such a case, the interacting null mutation will obliterate activity of one nonessential branch of the overall pathway, thereby partly compromising the activity of the entire pathway. From the point of view of the nonessential partner, the genetic enhancement must be parallel in nature as its local pathway is inactive. The genetic enhancement occurs because the hypomorphic mutation at the essential locus affects a parallel branch of the same overall pathway (Fig. 3.2).

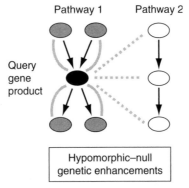

Figure 3.2. The functional relationships underlying hypomorphic–null interactions. Gene protein products (circles) act in two independent pathways that execute a common or buffering function. One pathway (2) is linear with the activity of each protein being fully dependent on the activity of the protein above it (indicated by black arrows) as for Fig. 3.1. The other pathway (1) is branched, with activity of the overall pathway being dependent only on the black, "query" hub protein. Null alleles in the genes encoding the gray proteins only abolish activity of the relevant branch of pathway 1. We consider the hypomorphic–null genetic enhancement interactions (gray lines) elicited by a "query" hypomorphic mutation affecting the gene that encodes the black protein. Genetic enhancements can occur between genes encoding components of the two pathways (dashed lines) or can also occur between genes encoding components of the same pathway 1 (filled gray lines).

Overall, we expect that hypomorphic–null interactions should enrich for "within pathway" relationships relative to null–null interactions, but not as strongly as do hypomorphic–hypomorphic interactions.

III. SYSTEMATIC DISCOVERY OF GENETIC ENHANCEMENTS

Experimental strategies for finding some genetic enhancements involving a gene of interest are well established in most model genetic organisms. However, large scale, unbiased, systematic identification of synthetic genetic interactions requires genome sequences and high throughput experimental protocols. Such protocols are thoroughly reviewed elsewhere and the interested reader is directed to consider these works (e.g., Boone et al., 2007). Here we focus on key features and strategies only.

A. Budding yeast *Saccharomyces cerevisiae*

Two strategies are available in budding yeast, synthetic genetic array (SGA) analysis and diploid synthetic lethal analysis by microarray (dSLAM) (Pan et al., 2004; Tong et al., 2001). In excess of 10,000 synthetic lethal interactions have been detected so far by these protocols and for hundreds of query genes.

1. SGA analysis

In this approach, an appropriate "query" haploid strain is generated containing a marked query mutation of interest (Tong et al., 2001). In the most common variant of SGA analysis, this haploid is mated on solid medium to an arrayed collection of marked null haploid mutants corresponding all ∼5000 nonessential yeast genes. The resulting and arrayed double heterozygotes are sporulated and an arrayed collection of haploid double mutant progeny of one mating type is directly selected. Failure of a double mutant colony to grow, or poor growth, at the appropriate grid position indicates a possible synthetic lethal interaction.

Any marked query mutation can be used. Most contain either a null allele of a nonessential gene or a hypomorphic allele of an essential gene (Tong et al., 2001, 2004). Indeed, the query allele can also be plasmid borne and expressed in *trans*, for example, a dominant negative allele expressed from a regulatable promoter (Krause et al., 2008).

2. dSLAM

Here, the query mutation is introduced into the collection of heterozygous diploids en mass in liquid culture by direct high efficiency transformation (Pan et al., 2004). The resulting mixed culture of heterozygotes is sporulated and double mutant haploid progeny of one mating type directly selected in liquid culture. The presence and growth of each double mutant in the mixed culture is quantified by microarray profiling that exploits the DNA barcodes specific to each null strain.

3. How the methods compare

Both SGA analysis and dSLAM are flexible and can accommodate any marked query allele. At present null–null and hypomorphic–null interactions can be readily identified by each method. Both methods are prone to both false positives and false negatives necessitating multiple repeats and independent confirmation of any suggested interactions.

Are the two methods equally effective? Some query mutations have been screened for synthetic lethal interactions using both SGA analysis and dSLAM (Pan et al., 2004). Direct comparison of the methods is thus possible. As a general rule, dSLAM analysis appears to identify significantly more inter-actions per query mutation than does SGA analysis and it does so with lower estimated false positive and false negative rates. As little as quarter of the confirmed interactions identified by dSLAM analysis may also be identified by SGA analysis. dSLAM thus appears to be more efficient and may be more sensitive than SGA analysis.

B. Fission yeast *Schizosaccharomyces pombe*

SGA analysis has recently and successfully been adapted for fission yeast by two different groups of workers (Dixon et al., 2008; Roguev et al., 2007). To date, a few dozen query genes have been screened against the viable null collection for synthetic lethal interactions.

C. *Escherichia coli*

SGA analysis has also been adapted for E. coli (Butland et al., 2008; Typas et al., 2008). A collection of marked haploid null mutants, each lacking a nonessential gene, has been generated in this bacterium. Although E. coli does not exist in a stable diploid form, chromosomal DNA from a donor strain can be transferred to a haploid recipient strain by conjugation and recombination. A marked query mutation can thus be crossed into arrayed null mutants and resulting arrayed double mutant haploid progeny analyzed. To date, only a small number of query mutations have been screened for synthetic lethal genetic interactions.

D. *Caenorhabditis elegans*

Two large-scale screens seeking genetic enhancements for some query gene mutations have been reported to date. Both exploit RNAi libraries to systematically downregulate expression of target genes in a given query mutant worm (Byrne *et al.*, 2007; Lehner *et al.*, 2006). Combinations of query mutation and RNAi targeting construct that cause a clear genetic enhancement are sought. The phenotypes screened include embryonic viability.

IV. NETWORK CHARACTERISTICS

A. Overall network architecture

For the budding yeast synthetic lethal genetic network, >95 of query genes tested take part in at least one synthetic lethal genetic interaction (Tong *et al.*, 2001, 2004). The available data for the *E. coli* and *S. pombe* genomes, although more limited in scale, are consistent with most genes in a microbe, be it eukaryotic or prokaryotic, being capable of participating in a synthetic lethal interaction in the vegetative state (Butland *et al.*, 2008; Dixon *et al.*, 2008; Typas *et al.*, 2008). Network coverage is wide.

The genetic enhancement network also appears to be very dense. On average, each budding yeast query gene elicits >30 synthetic genetic interactions (Tong *et al.*, 2001, 2004). Similar density has been indicated for the *E. coli* and *S. pombe* genomes (Butland *et al.*, 2008; Dixon *et al.*, 2008; Typas *et al.*, 2008).

The *C. elegans* genetic enhancement network is significantly less dense than those for its microbial cousins (Byrne *et al.*, 2007; Lehner *et al.*, 2006). The reason for this apparent low density if not well understood: it may be methodological; it may be a consequence of multicellularity and cell specialization.

All of the analyzed networks obey the power law: a small number of genes partake in lots of interactions and most genes partake in relatively few. For example, although the average density of the budding yeast network is ~30 synthetic lethal interactions per query gene tested, a small number of queries illicit many hundreds of interactions (Tong *et al.*, 2001, 2004). We discuss conservation of genetic enhancement networks later.

B. Functional basis for the observed interactions

1. Null–null interactions

We expect null–null genetic enhancements to reflect parallel, buffering relationships between the respective gene products. This expectation is confirmed by the data from large-scale networks, most clearly in the null–null dominated network analyzed in Tong *et al.* (2004). The majority of interactions involve genes pairs that do not share gene ontology (GO) annotations and appear to represent parallel, buffering relationships, many of which are surprising and may be very indirect (Tong *et al.*, 2004).

Such networks of interactions are not random. Genes with similar functions tend to share many common interaction partners—an example of network congruence that we return to later. Furthermore, interactions are more likely than random to occur between genes that share similar gene ontology annotations or whose products physically interact with each other. This enrichment is statistically significant but represents a small fraction of the entire set of interactions around query genes.

2. Hypomorphic–hypomorphic interactions

We expect hypomorphic–hypomorphic interactions to behave very differently to null–null interactions in enriching for "within pathway" interactions. Only one medium-scale analysis has directly sought such hypomorphic–hypomorphic interactions, in this case synthetic lethal interactions between pairs of essential budding yeast genes (Davierwala *et al.*, 2005). This analysis exploited titration expression alleles of essential genes. The ~1000 essential genes were reengineered *in situ* to contain a tetracycline-repressible operator sequence in their promoters. In the presence of the tetracycline binding protein and tetracycline, transcription from such alleles should be repressed to some extent. Most of the resulting alleles are viable, indicating that residual expression is sufficient to support a minimal level of biological activity (Mnaimneh *et al.*, 2004).

Some titration expression alleles were used as queries and screened against a limited collection of other, marked titration expression alleles. The resulting network is dense and again appears to obey the power law (Davierwala *et al.*, 2005). However, the interactions do not appear qualitatively different from null–null interactions (Davierwala *et al.*, 2005) as represented by the null–null dominated Tong *et al.* (2004) network. Both networks show similar low levels of GO co-annotation between interacting genes. Furthermore, the chances of a genetic interaction being accompanied by a physical interaction between the gene products is also similar in the two networks, even though such "physical-genetic" interactions are thought to be robust indicators of the degree of "within pathway" interactions in a network (Tong *et al.*, 2004).

These findings do not support the expectation that genetic enhancement interactions between hypomorphic alleles should be enriched for, if not dominated by, "within pathway" interactions. Our understanding of genetic interactions may be wrong. Alternatively, many titration expression alleles may not be truly hypomorphic. Comprehensive, side-by-side comparisons between titration expression and known hypomorphic alleles have not been made for the same genes. The jury is out. The emergence of alternative mutant collections constructed to generate hypomorphic alleles such as temperature sensitive, DAmP and degron alleles (Ben-Aroya *et al.*, 2008; Dohmen and Varshavsky, 2005; Yan *et al.*, 2008) will help establish the true nature of networks involving hypomorphic alleles. However, some limited results available to date hint at the true nature of hypomorphic–hypomorphic networks. We discuss one example here involving nonallelic noncomplementation (NNC) and the *ACT1* gene.

The essential *ACT1* gene of budding yeast encodes actin, the building block for the actin cytoskeleton. Heterozygous null diploids lacking one copy of the *ACT1* gene are viable but display phenotypes indicative of some degree of haploinsufficiency (Haarer *et al.*, 2007). These heterozygotes thus behave as being hypomorphic for Act1 function.

A recent analysis sought NNC interactions between the *act1Δ* mutation and null mutations in any of the nonessential yeast genes (Haarer *et al.*, 2007). Such a genetic interaction occurs when the double heterozygote is inviable/slow growing. Because the interaction occurs between two null mutations each in the heterozygous state, it should correspond to a hypomorphic–hypomorphic synthetic lethal interaction, but in this case, between an essential and a nonessential gene.

If the genetic enhancement network around *ACT1* is enriched for "within pathway" interactions, then many of the interacting gene products should play a role in the actin cytoskeleton. Encouragingly, two thirds of the null haploid mutants corresponding to the interacting genes display actin defects on their own (Haarer *et al.*, 2007). The hypomorphic–hypomorphic genetic enhancement interaction network around the *ACT1* gene appears to be dominated by "within pathway" interactions.

3. Hypomorphic–null interactions

Hypomorphic–null interactions should also enrich for "within pathway" interactions. Data from screens involving titration expression alleles are again not supportive of any such enrichment over a null–null dominated network (Davierwala *et al.*, 2005). However, titration expression alleles are not necessarily representative of how hypomorphic alleles behave.

Some genuinely hypomorphic alleles of essential yeast query genes have been screened for hypomorphic–null synthetic lethal interactions. Here, we discuss one case briefly and go on to consider many of these cases *en mass* in Section IV.B.4.

The budding yeast protein kinase C, encoded by the *PKC1* gene, is an essential hub protein in the cell wall integrity pathway that is required for ordered expansion of the cell surface (reviewed in Levin, 2005). Multiple nonessential branches feed into and out of a core module containing Pkc1. We recently identified the synthetic lethal network of nonessential genes around *PKC1*, using a dominant negative allele of the gene expressed in trans and using SGA analysis methodology (Krause *et al.*, 2008). Of the 21 interactions, we found evidence in support of 14, two thirds, being "within pathway" interactions, with seven involving established components of the Pkc1 pathway. Such enrichment for "within pathway" interactions is not an unusual property of genes involved in Pkc1 signaling: the null–null network around *SLT2*, encoding a key downstream pathway component is largely distinct from the *PKC1* network (Krause and Gray, 2009).

4. Directly comparing hypomorphic–null and null–null subnetworks

Many hypomorphic alleles of essential genes have been screened for hypomorphic–null synthetic lethal interactions in budding yeast over the last 5 years, most of which are temperature-sensitive alleles. We surveyed the literature and identified 26 such cases (Krause, Ding, and Gray, unpublished work), where the networks had all been determined using the same SGA analysis approach (Audhya *et al.*, 2004; Budd *et al.*, 2005; Kozminski *et al.*, 2003; Measday *et al.*, 2005; Montpetit *et al.*, 2005; Parsons *et al.*, 2004; Sarin *et al.*, 2004; Sciorra *et al.*, 2005; Suter *et al.*, 2004; Tong *et al.*, 2001, 2004). These 26 essential queries elicited 792 synthetic lethal interactions, an average of ~ 30 per query mutation (Krause, Ding, and Gray, unpublished work), similar to that density noted for null–null networks (Tong *et al.*, 2004). We also generated a reference set of ~ 2500 null–null synthetic lethal interactions identified in Tong *et al.* (2004). We compared the frequencies of GO term co-annotation and "physical-genetic" interactions in the hypomorphic–null versus null–null networks (Krause, Ding, and Gray, unpublished work). As shown in Fig. 3.3, we find that all the key indicators (save for localization including cytoplasm) are significantly enriched in the hypomorphic–null set. The most striking finding is the sevenfold enrichment for physical-genetic interactions in the hypomorphic–null network. The absolute frequency of such "physical-genetic" interactions in the hypomorphic–null network approaches the frequency noted for gene pairs predicted to perform the most similar *in vivo* functions (Tong *et al.*, 2004). Although this analysis is preliminary, it strongly supports the possibility that hypomorphic alleles indeed enrich for "within pathway" interactions.

Our findings indicate that titration expression and hypomorphic alleles of essential genes behave very differently in genetic enhancement networks. The nature of the network is thus surprisingly allele dependent.

Finally, the large-scale Tong *et al.* (2004) network is often used as the reference network and is treated as being uniform in nature. This network is indeed dominated by null–null interactions between nonessential genes, but

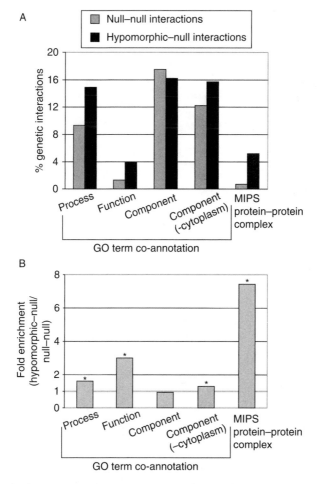

Figure 3.3. The frequency of GO term co-annotation and parallel protein–protein interactions ("physical-genetic" interactions) differs between sample null–null and hypomorphic–null subnetworks (792 and 2721 interactions, respectively) of the budding yeast synthetic lethal network. (A) The absolute frequencies of GO term co-annotation (process, function, component, and component (-cytoplasm)) and parallel MIPS protein–protein interactions are shown for both the null–null and hypomorphic–null interaction subnetworks. (B) The fold enrichments in frequency of GO term co-annotation and parallel protein–protein interactions in the hypomorphic–null set relative to their frequencies in the null–null set are shown.

the network is mixed. It also contains a small but significant fraction of hypomorphic–null interactions elicited by some essential query genes. Indeed, this small fraction of interactions contributed disproportionately to some characteristics of the overall Tong *et al.* network, most especially, the frequency of physical–genetic interactions (Krause, Ding, and Gray, unpublished work).

Our analysis points to the danger in interpreting genetic interactions when the nature of the interacting alleles is not well understood. Allele type should not be ignored and may prove a key in properly interpreting and exploiting genetic enhancement networks.

V. EXTRACTING FUNCTIONAL RELATIONSHIPS FROM GLOBAL NETWORKS

Functional relationships between gene products can be inferred from networks of genetic enhancements. Before we summarize the main strategies for extracting these relationships, it should be noted that the analytical power of genetic enhancement networks is further enhanced when additional genetic interactions such as genetic suppression interactions are also included in the analysis, as indicated by epistasis mapping (E-MAP) analyses performed to date for small numbers of closely related genes (e.g., see Schuldiner *et al.*, 2005; Collins *et al.*, 2007). However, genetic enhancements dominate the extant genetic interaction and here, we focus on exploiting these networks.

Pairwise null–null synthetic genetic interactions are underpinned by parallel functional relationships. Hence, the network around a query gene of interest does not itself directly inform the function of the query gene product. However, genes that share the same interactions are likely to function closely together (Gray and Krause, 2007; Tong *et al.*, 2001). Such congruence between networks is a key to identifying functionally related gene pairs but requires significant extant knowledge of the relevant portion of the global synthetic genetic network.

Tong *et al.* (2004) noted that genes that the most similar pattern of synthetic lethal interactions tend to be functionally related. Such gene products tend to co-localize, exist in the same physical complex, and share other GO term annotations. Other workers have extended this observation and demonstrated that congruence, in number or pattern of interactions or both, is highly indicative of close functional relationships between gene products (Ye *et al.*, 2005).

Consider two nonessential genes whose products act in a single linear pathway. In addition to the synthetic genetic networks around the genes being highly congruent, we would also expect that the two genes do not interact with each other: the local pathway is inactivated by each single null mutation.

Although simplistic, strict application of this model can identify, with surprising specificity, functional pathways in the living cell. For example, we previously showed that starting with the nonessential chitin synthase genes of yeast, we could correctly identify components that synthesize, deliver, localize, or activate different isozymes within the cell (Gray and Krause, 2007).

However, most *in vivo* pathways are not linear. The presence of a null–null synthetic genetic interaction between two genes does not necessarily preclude their gene products acting in the same, branched pathway or process. Other workers have considered the more general case where pathway architecture, and the very concept of a functional pathway, is more difficult to define. In this case, a pathway can still be considered as a group of genes with highly congruent genetic enhancement networks but with few genetic interactions occurring between them or whose products tend to physically associate, or both (Ye *et al.*, 2005).

Hypomorphic alleles appear to enrich for "within pathway" interactions, where the network directly identifies components of the same pathway. In seeking components of the same pathway, it may thus be possible to productively study the network around a given query gene without reference to the more global genetic interaction network for the organism. The *ACT1* and *PKC1* networks in budding yeast are good examples and have been discussed earlier.

However, network congruence also applies here. We expect high congruence between the networks of functionally related essential genes. However, this prediction only holds when like networks are compared: null–null versus null–null, hypomorphic–null versus hypomorphic–null, and so on. It is more difficult to predict how different classes of network compare. For example, if networks around hypomorphic alleles are predominately "within pathway," then there will be little congruence noted between the hypomorphic–null and null–null networks around an essential and a nonessential gene, respectively, whose products act in the same functional pathway. In contrast, if many synthetic genetic interactions involving hypomorphic alleles are underpinned by parallel functional relationships, then congruence can be noted between these networks.

Recent large-scale analysis of chemical sensitivity finds some measurable phenotype change for 97% of yeast null mutants, in the heterozygous, or homozygous states, or both (Hillenmeyer *et al.*, 2008). Many of these observed chemical sensitivities may correspond to underlying synthetic genetic interactions (Parsons *et al.*, 2004, 2006) between the affected genes. As such, many genes could be grouped into functional clusters by virtue of congruence between their networks of chemical sensitivities.

VI. CONSERVATION

Many genes and their functions are highly conserved across eukaryotes. Are genetic interactions between orthologous gene pairs also conserved between species? If so, then the genetic interaction network determined in one experimentally tractable species could be provisionally mapped onto the genome of any other species, greatly assisting and accelerating the understanding of many complex genetic traits in less experimentally tractable organisms. Indeed, could the experimentally determined genetic interaction network of a simple organism like yeast be used to predict with some robustness the genetic modifiers of many inherited human disease genes?

Unfortunately, comparative analysis between genetic enhancement networks in different species is severely limited by the amount of relevant, available, and systematically determined data. Some data are available on the overlap between the synthetic genetic networks of the budding and fission yeasts and on the nature of the worm network (Byrne *et al.*, 2007; Dixon *et al.*, 2008; Lehner *et al.*, 2006; Pan *et al.*, 2004; Tischler *et al.*, 2008; Tong *et al.*, 2001, 2004). Here, we focus on the overlap between the yeasts as the datasets are most robust and determined by similar methods.

Although the budding and fission yeasts are both fungi, and indeed ascomycetes, the divergence of their lineages is thought to be relatively ancient. Occurring some 300–400 million years ago (Sipiczki 2000). Analysis of the synthetic lethal networks of these two species is thus a relatively robust test of how strongly synthetic genetic networks are conserved.

The available data for fission yeast can be meaningfully compared to those of budding yeast because orthologous gene sets were used as queries in both species and the networks determined using similar protocols. Although the available data on fission yeast are limited in extent, overlap between the networks of the two yeasts has been observed (Dixon *et al.*, 2008; Roguev *et al.* 2008). Specifically, approximately a quarter of the interactions identified in fission yeast by screening query mutations against the null mutant collection occur between gene pairs that have orthologs in budding yeast. These interactions are thus conserved between the two species.

A more focused analysis of genetic interactions has been reported for genes involved in chromosome biology in fission yeast (Roguev *et al.*, 2008). This analysis focused on detecting genetic interactions, both suppressions and enhancements, between a small number of *pombe* genes (epistasis mapping or E-MAP analysis), most of which have orthologs in budding yeast. The conservation of genetic enhancements between the yeast networks was again relatively low at ~17%, consistent with Dixon *et al.* (2008). Interestingly, interactions tended to occur between genes encoding components of the same molecular complexes in both species, even if the specific interactions tended to be different.

These results point to some significant conservation of the genetic enhancement network between species. How robust is this conclusion? We consider the pessimistic and optimistic interpretations in turn.

We begin with the pessimist's view. First, approximately a quarter of the synthetic lethal interactions between orthologous gene pairs are conserved between the two species. This number overestimates the extent of overall network overlap since many interactions in the genome-wide networks will occur with or between genes that have no orthologs in the other species. It is already clear from the available data that orthologs do not genetically interact only with other orthologs. Given that genes with orthologs in the yeasts comprise a minority of genes in higher plants and animals, the yeast genetic interaction networks are likely to at best correspond to only a small fraction of the overall genetic enhancement networks in these species. Second, the known yeast networks are dominated by interactions involving null alleles of nonessential genes. However, null alleles are not necessarily representative of the alleles that underlie genetic interactions occurring in natural populations of any species. Third, the two yeasts have similar simple lifestyles and live in similar ecological niches. One might expect extrapolation of such interaction networks to multicellular organisms with cell specialization and organ systems to be more challenging.

Now, the optimist's response. First, the reported overlap between the synthetic lethal networks of the two yeasts may grossly underestimate the true extent of overlap. One possible reason lies in the sub-saturating nature of SGA screens. In budding yeast, dSLAM analysis detects *bona fide* interactions significantly more efficiently than does SGA analysis (Pan *et al.*, 2004). If we consider the results of dSLAM to identify the true synthetic lethal network of budding yeast, then we can estimate SGA to be ~50% effective in identifying these interactions. We do not know the basis for this relative inefficiency but there is no reason to expect it to also be evolutionarily conserved. If the SGA analysis of budding and fission yeast identified ~50% of the true interactions and the networks are fully conserved, one would expect to observe a mere 50% overlap between interactions involving orthologous gene pairs. The methodology used to determine the yeast networks may have obscured the true extent of overlap. Second, a genetic enhancement may be conserved between the two yeast species but differ in magnitude or even in the affected phenotype because of the different genomic contexts.

The limited data available for the genetic enhancement network of the worm support poor conservation of genetic networks between species. The observed genetic enhancement network of C. *elegans* is much less dense than those of the yeasts (Byrne *et al.*, 2007; Lehner *et al.*, 2006; Tischler *et al.*, 2008) with few shared interactions (Tischler *et al.*, 2008). However, some caution is again advised in interpreting these limited data. First, the worm interactions were detected using RNAi screening methodology. RNAi treatment is thought to reduce but not to abolish expression of the target gene and thus resembles a hypomorphic mutation (Perrimon and Mathey-Pervot, 2007). As we have discussed earlier, genetic interactions may be very allele dependent. It may be inappropriate to compare the null–null dominated networks of the yeasts with the worm network. Second, the worm and yeast screens were performed by different experimental approaches. As is demonstrated by SGA analysis and dSLAM in the budding yeast, methodological differences can greatly affect the outcome of genetic interaction screens (Pan et al., 2004). Third, RNAi screening may not be as robust as hoped. For example, recent RNAi-based screens seeking genes that inhibit lifespan identified large numbers of positives (Curran and Ruvkun, 2007; Hamilton et al., 2005; Hansen et al., 2005; Lee et al., 2003). However, the datasets overlapped by <15%. In contrast, genetic interaction screens in the yeasts are performed multiple times and all interactions are independently confirmed. The lack of overlap between the worm and yeast interaction networks could thus be due to discordance in the quality and robustness of the datasets being compared.

At present, comparing the genetic interaction networks of different species is fraught with danger and complication. The jury is out on the extent and utility of conservation of genetic interaction networks.

VII. CONCLUSIONS

We are in a golden era for understanding and exploiting genetic interactions. Fortuitously, this blooming in fundamental genetics is coinciding with the emerging need to understand and exploit multigenic traits in crop plants, farm animals, and humans.

The technology for systematically identifying genetic interactions is well advanced in single celled microbes, both prokaryote and eukaryote. The synthetic lethal network of the budding yeast is the best understood and will doubtless be the first network of any organism to be fully characterized in the next few years.

Uncovering the networks of model microbes will contribute greatly to our understanding of functional genomics, with the possibility of our improved understanding of gene function being immediately transferred to orthologous genes in other organisms. These networks will also clarify the rules underpinning genetic interactions, rules that may themselves be universal. These networks reflect in vivo functional relationships between gene products and must be a key input into any understanding of cellular life as a system.

Some key features of genetic interaction networks are less well understood or underexploited. The allele dependence of genetic interactions strikes us as key to understanding genetic networks in general. Other confounding factors, such as phenotype, environment, and cell state, may also greatly affect whether a genetic interaction is observed or not. Furthermore, the terminal phenotype associated with a genetic enhancement can greatly inform the biological role of the interacting gene products.

We are beginning to explore the extent to which genetic interaction networks are conserved between species and kingdoms. Preliminary data point to general features of networks being conserved, but it is less clear if individual interactions are also strongly conserved. Clarifying the extent of network conservation will be a key to understanding how best to understand the genetics underlying many commercially and clinically important multigenic traits.

Acknowledgments

This work was supported by research grants from the Biotechnology and Biological Sciences Research Council (C/20144) to J. V. G. We thank members of the lab and colleagues for helpful discussions and Huiming Ding for his help in preparing the data shown in Fig. 3.3.

References

Altenburg, E., and Muller, H. J. (1920). The genetic basis of truncate wing, an inconstant and modifiable character in *Drosophila*. *Genetics* **5,** 1–59.

Audhya, A., Leowith, R., Parsons, A. B., Gao, L., Tabuchi, M., Zhou, H., Boone, C., Hall, M. N., and Emr, S. D. (2004). Genome-wide lethality screen identifies new PI4, 5P2 effectors that regulate the actin cytoskeleton. *EMBO J.* **23,** 3747–3757.

Ben-Aroya, S., Coombes, C., Kwok, T., O'Donnell, K. A., Boeke, J. D., and Hieter, P. (2008). Toward a comprehensive temperature-sensitive mutant repository of the essential genes of *Saccharomyces cerevisiae*. *Mol. Cell* **30,** 248–258.

Boone, C., Bussey, H., and Andrews, B. J. (2007). Exploring genetic interactions and networks with yeast. *Nat. Rev. Genet.* **8,** 437–449.

Budd, M. E., Tong, A. H., Polaczek, P., Peng, X., Boone, C., and Campbell, J. L. (2005). A network of multi-tasking proteins at the DNA replication fork preserves genome stability. *PLoS Genet* **1,** e61.

Butland, G., Babu, M., Díaz-Mejía, J. J., Bohdana, F., Phanse, S., Gold, B., Yang, W., Li, J., Gagarinova, A. G., Pogoutse, O., Mori, H., Wanner, B. L., *et al.* (2008). eSGA: E. coli synthetic genetic array analysis. *Nat. Methods* **5,** 789–795.

Byrne, A. B., Weirauch, M. T., Wong, V., Koeva, M., Dixon, S. J., Stuart, J. M., and Roy, P. J. (2007). A global analysis of genetic interactions in *Caenorhabditis elegans*. *J. Biol.* **6,** 8.

Collins, S. R., Miller, K. M., Maas, N. L., Roguev, A., Fillingham, J., Chu, C. S., Schuldiner, M., Gebbia, M., Recht, J., Shales, M., Ding, H., Xu, H., *et al.* (2007). Functional dissection of protein complexes involved in yeast chromosome biology using a genetic interaction map. *Nature* **446,** 806–810.

Curran, S. P., and Ruvkun, G. (2007). Lifespan regulation by evolutionarily conserved genes essential for viability. *PLoS Genet.* **3,** e56.

Davierwala, A. P., Haynes, J., Li, Z., Brost, R. L., Robinson, M. D., Yu, L., Mnaimneh, S., Ding, H., Zhu, H., Chen, Y., Cheng, X., Brown, G. W., *et al.* (2005). The synthetic genetic interaction spectrum of essential genes. *Nat. Genet* **37,** 1147–1152.

Dixon, S. J., Fedyshyn, Y., Koh, J. L., Prasad, T. S., Chahwan, C., Chua, G., Toufighi, K., Baryshnikova, A., Hayles, J., Hoe, K. L., Kim, D. U., Park, H. O., et al. (2008). Significant conservation of synthetic lethal genetic interaction networks between distantly related eukaryotes. Proc. Natl. Acad. Sci. USA 105, 16653–16658.

Dohmen, R. J., and Varshavsky, A. (2005). Heat-inducible degron and the making of conditional mutants. Methods Enzymol. 399, 799–822.

Gray, J. V., and Krause, S. A. (2007). Identifying in vivo pathways using genome-wide genetic networks. Biochem. Soc. Trans. 35, 1538–1541.

Gray, J. V., Ogas, J. P., Kamada, Y., Stone, M., Levin, D. E., and Herskowitz, I. (1997). A role for the Pkc1 MAP kinase pathway of Saccharomyces cerevisiae in bud emergence and identification of a putative upstream regulator. EMBO J. 16, 4924–4937.

Guarente, L. (1993). Synthetic enhancement in gene interaction: A genetic tool come of age. Trends Genet. 9, 362–366.

Haarer, B., Viggiano, S., Hibbs, M. A., Troyanskaya, O. G., and Amberg, D. C. (2007). Modeling complex genetic interactions in a simple eukaryotic genome: Actin displays a rich spectrum of complex haploinsufficiencies. Genes Dev. 21, 148–159.

Hamilton, B., Dong, Y., Shindo, M., Liu, W., Odell, I., Ruvkun, G., and Lee, S. S. (2005). A systematic RNAi screen for longevity genes in C. elegans. Genes. Dev. 19, 1544–1555.

Hansen, M., Hsu, A. L., Dillin, A., and Kenyon, C. (2005). New genes tied to endocrine, metabolic, and dietary regulation of lifespan from a Caenorhabditis elegans genomic RNAi screen. PLoS Genet. 1, 119–128.

Hillenmeyer, M. E., Fung, E., Wildenhain, J., Pierce, S. E., Hoon, S., Lee, W., Proctor, M., St Onge, R. P., Tyers, M., Koller, D., Altman, R. B., Davis, R. W., et al. (2008). The chemical genomic portrait of yeast: Uncovering a phenotype for all genes. Science 320, 362–365.

Hartman, J. L., IV, Garvik, B., and Hartwell, L. (2001). Principles for the buffering of genetic variation. Science 291, 1001–1004.

Jeong, H., Mason, S. P., Barabási, A. L., and Oltvai, Z. N. (2001). Lethality and centrality in protein networks. Nature 411, 41–42.

Koch, C., Schleiffer, A., Ammerer, G., and Nasmyth, K. (1996). Switching transcription on and off during the yeast cell cycle: Cln/Cdc28 kinases activate bound transcription factor SBF (Swi4/Swi6) at start, whereas Clb/Cdc28 kinases displace it from the promoter in G2. Genes Dev. 10, 129–141.

Kozminski, K. G., Beven, L., Angerman, E., Tong, A. H., Boone, C., and Park, H. O. (2003). Interaction between a Ras and a Rho GTPase couples selection of a growth site to the development of cell polarity in yeast. Mol. Biol. Cell 14, 4958–4970.

Krause, S. A., and Gray, J. V. (2009). The functional relationships underlying a synthetic genetic network. Commun. Integ. Biol. 2, 4–6.

Krause, S. A., Xu, H., and Gray, J. V. (2008). The synthetic genetic network around PKC1 identifies novel modulators and components of protein kinase C signaling in Saccharomyces cerevisiae. Eukaryot. Cell 7, 1880–1887.

Lee, S. S., Lee, R. Y., Fraser, A. G., Kamath, R. S., Ahringer, J., and Ruvkun, G. (2003). A systematic RNAi screen identifies a critical role for mitochondria in C. elegans longevity. Nat. Genet. 33, 40–48.

Lehner, B., Crombie, C., Tischler, J., Fortunato, A., and Fraser, A. G. (2006). Systematic mapping of genetic interactions in C. elegans. Nat. Genet. 38, 896–903.

Levin, D. E. (2005). Cell wall integrity signaling in Saccharomyces cerevisiae. Microbiol. Mol. Biol. Rev. 69, 262–291.

Madden, K., Sheu, Y. J., Baetz, K., Andrews, B., and Snyder, M. (1997). SBF cell cycle regulator as a target of the yeast PKC-MAP kinase pathway. Science 275, 1781–1784.

Measday, V., Baetz, K., Guzzo, J., Yuen, K., Kwok, T., Sheikh, B., Ding, H., Ueta, R., Hoac, T., Cheng, B., Pot, I., Tong, A., *et al.* (2005). Systematic yeast synthetic lethal and synthetic dosage lethal screens identify genes required for chromosome segregation. *Proc. Natl. Acad. Sci. USA* **102**, 13956–13961.

Mnaimneh, S., Davierwala, A. P., Haynes, J., Moffat, J., Peng, W. T., Zhang, W., Yang, X., Pootoolal, J., Chua, G., Lopez, A., Trochesset, M., Morse, D., *et al.* (2004). Exploration of essential gene functions via titratable promoter alleles. *Cell* **118**, 31–44.

Montpetit, B., Thorne, K., Barrett, I., Andrews, K., Jadusingh, R., Hieter, P., and Measday, V. (2005). Genome-wide synthetic lethal screens identify an interaction between the nuclear envelope protein, Apq12p, and the kinetochore in *Saccharomyces cerevisiae*. *Genetics* **171**, 489–501.

Ooi, S. L., Pan, X., Peyser, B. D., Ye, P., Meluh, P. B., Yuan, D. S., Irizarry, R. A., Bader, J. S., Spencer, F. A., and Boeke, J. D. (2006). Global synthetic-lethality analysis and yeast functional profiling. *Trends Genet.* **22**, 56–63.

Pan, X., Yuan, D. S., Xiang, D., Wang, X., Sookhai-Mahadeo, S., Bader, J. S., Hieter, P., Spencer, F., and Boeke, J. D. (2004). A robust toolkit for functional profiling of the yeast genome. *Mol. Cell* **16**, 487–496.

Parsons, A. B., Brost, R. L., Ding, H., Li, Z., Zhang, C., Sheikh, B., Brown, G. W., Kane, P. M., Hughes, T. R., Boone, C., *et al.* (2004). Integration of chemical-genetic and genetic interaction data links bioactive compounds to cellular target pathways. *Nat. Biotechnol.* **22**, 62–69.

Parsons, A. B., Lopez, A., Givoni, I. E., Williams, D. E., Gray, C. A., Porter, J, Chua, G., Sopko, R., Brost, R. L., Ho, C. H., Wang, J., Ketela, T., *et al.* (2006). Exploring the mode-of-action of bioactive compounds by chemical-genetic profiling in yeast. *Cell* **126**, 611–625.

Perrimon, N., and Mathey-Prevot, B. (2007). Applications of high-throughput RNA interference screens to problems in cell and developmental biology. *Genetics* **175**, 7–16.

Roguev, A., Wiren, M., Weissman, J. S., and Krogan, N. J. (2007). High-throughput genetic interaction mapping in the fission yeast *Schizosaccharomyces pombe*. *Nat. Methods* **4**(10), 861–866.

Roguev, A., Bandyopadhyay, S., Zofall, M., Zhang, K., Fischer, T., Collins, S. R., Qu, H., Shales, M., Park, H. O., Hayles, J., Hoe, K. L., Kim, D. U., *et al.* (2008). Conservation and rewiring of functional modules revealed by an epistasis map in fission yeast. *Science* **322**, 405–410.

Sarin, S., Ross, K. E., Boucher, L., Green, Y., Tyers, M., and Cohen-Fix, O. (2004). Uncovering novel cell cycle players through the inactivation of securin in budding yeast. *Genetics* **168**, 1763–1771.

Schuldiner, M., Collins, S. R., Thompson, N. J., Denic, V., Bhamidipati, A., Punna, T., Ihmels, J., Andrews, B., Boone, C., Greenblatt, J. F., Weissman, J. S., and Krogan, N. J. (2005). Exploration of the function and organization of the yeast early secretory pathway through an epistatic miniarray profile. *Cell* **123**, 507–519.

Sciorra, V. A., Audhya, A., Parsons, A. B., Segev, N., Boone, C., and Emr, S. D. (2005). Synthetic genetic array analysis of the PtdIns 4-kinase Pik1p identifies components in a Golgi-specific Ypt31/rab-GTPase signaling pathway. *Mol. Biol. Cell* **16**, 776–793.

Sipiczki, M (2000). Where does fission yeast sit on the tree of life? *Genome Biol.* **1**, 1011.1–1011.4, reviews.

St Onge, R. P., Mani, R., Oh, J., Proctor, M., Fung, E., Davis, R. W., Nislow, C., Roth, F. P., and Giaever, G. (2007). Systematic pathway analysis using high-resolution fitness profiling of combinatorial gene deletions. *Nat. Genet.* **39**, 199–206.

St Johnston, D. (2002). The art and design of genetic screens: *Drosophila melanogaster*. *Nat. Genet.* **3**, 176–188.

Suter, B., Tong, A., Chang, M., Yu, L., Brown, G. W., Boone, C., and Rine, J. (2004). The origin recognition complex links replication, sister chromatid cohesion and transcriptional silencing in *Saccharomyces cerevisiae*. *Genetics* **167**, 579–591.

Tischler, J., Lehner, B., and Fraser, A. G. (2008). Evolutionary plasticity of genetic interaction networks. *Nat. Genet.* **40,** 390–391.

Tong, A. H., Evangelista, M., Parsons, A. B., Xu, H., Bader, G. D., Pagé, N., Robinson, M., Raghibizadeh, S., Hogue, C. W., Bussey, H., Andrews, B., Tyers, M., et al. (2001). Systematic genetic analysis with ordered arrays of yeast deletion mutants. *Science* **294,** 2364–2368.

Tong, A. H., Lesage, G., Bader, G. D., Ding, H., Xu, H., Xin, X., Young, J., Berriz, G. F., Brost, R. L., Chang, M., Chen, Y., Cheng, X., et al. (2004). Global mapping of the yeast genetic interaction network. *Science* **303,** 808–813.

Typas, A., Nichols, R. J., Siegele, D. A., Shales, M., Collins, S. R., Lim, B., Braberg, H., Yamamoto, N., Takeuchi, R., Wanner, B. L., Mori, H., Weissman, J. S., et al. (2008). High-throughput, quantitative analyses of genetic interactions in *E. coli. Nat. Methods* **5,** 781–787.

Watanabe, Y., Takaesu, G., Hagiwara, M., Irie, K., and Matsumoto, K. (1997). Characterization of a serum response factor-like protein in *Saccharomyces cerevisiae*, Rlm1, which has transcriptional activity regulated by the Mpk1 (Slt2) mitogen-activated protein kinase pathway. *Mol. Cell Biol.* **17,** 2615–2623.

Winzeler, E. A., Shoemaker, D. D., Astromoff, A., Liang, H., Anderson, K., Andre, B., Bangham, R., Benito, R., Boeke, J. D., Bussey, H., Chu, A. M., Connelly, C., et al. (1999). Functional characterization of the *S. cerevisiae* genome by gene deletion and parallel analysis. *Science* **285,** 901–906.

Yan, Z., Costanzo, M., Heisler, L. E., Paw, J., Kaper, F., Andrews, B. J., Boone, C., Giaever, G., and Nislow, C. (2008). Yeast Barcoders: A chemogenomic application of a universal donor-strain collection carrying bar-code identifiers. *Nat. Methods* **5,** 719–725.

Ye, P., Peyser, B. D., Pan, X., Boeke, J. D., Spencer, F. A., and Bader, J. S. (2005). Gene function prediction from congruent synthetic lethal interactions in yeast. *Mol. Syst. Biol.* **1,** 2005.0026.

Index